西安交通大学 "十三五"规划教材

出国访学准备
跨文化学术交流基础

主 编 牛 莉
副主编 龚 颖 吴 萍
编 者 胡 洁 楚建伟 成 旻

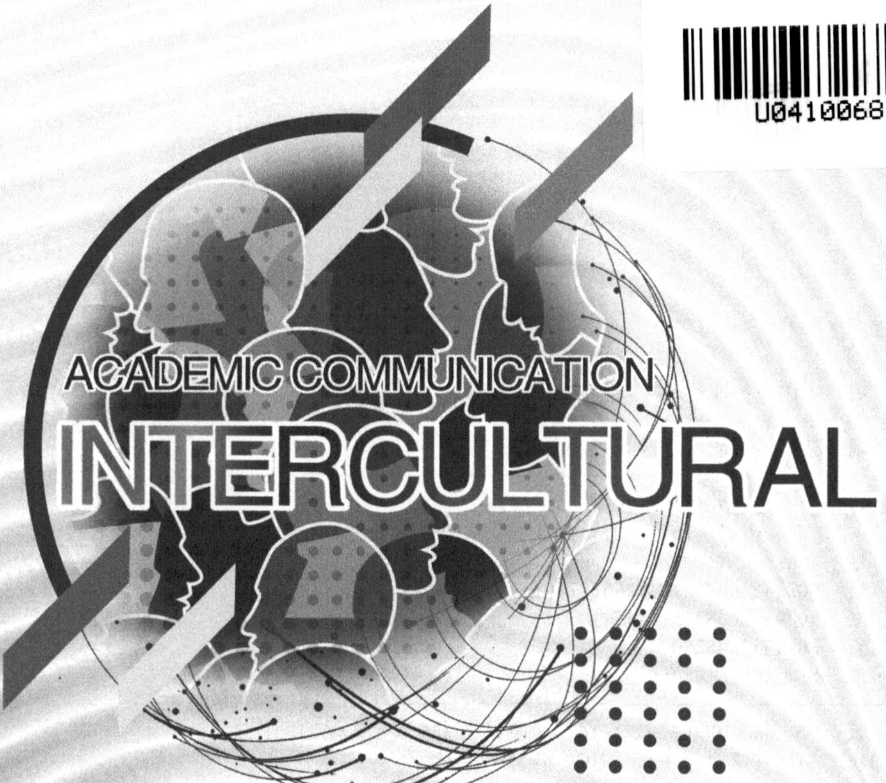

西安交通大学出版社
XI'AN JIAOTONG UNIVERSITY PRESS
国家一级出版社
全国百佳图书出版单位

图书在版编目(CIP)数据

出国访学准备：跨文化学术交流基础：英文 / 牛莉主编.
—西安：西安交通大学出版社，2021.2(2022.3 重印)
ISBN 978-7-5693-1952-1

Ⅰ.①出… Ⅱ.①牛… Ⅲ.①学术交流-研究-英文
Ⅳ.①G321.5

中国版本图书馆 CIP 数据核字(2020)第 249953 号

书　　名　出国访学准备：跨文化学术交流基础
主　　编　牛　莉
责任编辑　蔡乐芊

出版发行　西安交通大学出版社
　　　　　(西安市兴庆南路 1 号　邮政编码 710048)
网　　址　http://www.xjtupress.com
电　　话　(029)82668357　82667874(市场营销中心)
　　　　　(029)82668315(总编办)
传　　真　(029)82668280
印　　刷　西安五星印刷有限公司

开　　本　720 mm×1000 mm　1/16　印张 9.375　字数 212 千字
版次印次　2021 年 2 月第 1 版　2022 年 3 月第 2 次印刷
书　　号　ISBN 978-7-5693-1952-1
定　　价　40.00 元

读者购书、书店添货，如发现印装质量问题，请与本社市场营销中心联系调换。
订购热线：(029)82665248　(029)82665249
投稿热线：(029)82665371
读者信箱：xjtu_rw@163.com

前 言

《出国访学准备:跨文化学术交流基础》是西安交通大学“十三五”规划教材,入选西安交通大学名课、名教材建设项目。本教材编写旨在帮助有出国访学或交流需求的学生做好相关学术英语能力、跨文化交技能力和心理等方面的准备,更重要的是培养学生跨文化学术交流的能力,为他们成为国际化人才打好基础。

本教材包括 4 个准备模块,共 6 个部分:

模块 1:申请准备,包括申请文书(Personal Statement and CV)的写作和备考国际英语测试(IELTS, TOEFL and GRE)中的难点部分,即写作任务。详见本教材 Section 1 和 Section 2。

模块 2: 学业技能准备,包括对学生在撰写课程论文、实验报告、参与项目研究过程中所需的文献阅读、综述及其他学术读写技能进行训练。同时,针对培养学生学术诚信价值观念,教材选用了 APA 格式对学生严格遵守论文引述及参考文献格式进行系统训练。详见本教材 Section 3 和 Section 4。

模块 3:学术交流准备,采用多种常见的学术口语交流形式,如研讨会(seminar)、学术演讲(presentations)、专题讨论会(panel discussion)等,结合相关教学内容,培养学生学术交流表达能力。详见本教材 Section 5。

模块 4:跨文化交际能力准备,学生通过对文化定义及文化元素的深入学习和研究分析,了解跨文化交际中存在的问题及解决方法。同时,引导学生学习如何讲好中国故事,在国际交流中建立文化自信,做好文化沟通和理解的桥梁。详见本教材 Section 6。

本教材设计采用“概念驱动”理论,从而保证了“以学为中心”教学理念的实现。章节编写以掌握相关能力为目标,通过完成不同的任务,使学生在“做”中学习知识并掌握技能。本教材从 2016 年 9 月开始试用至今,不断打磨,配合课程学习,学生普遍反映学到了有用的知识和技能。

最后,特别感谢为本书编写做出巨大贡献的编者们和为教材建设提出宝贵建议的同学们!

牛 莉

2021 年 2 月

Map of Learning

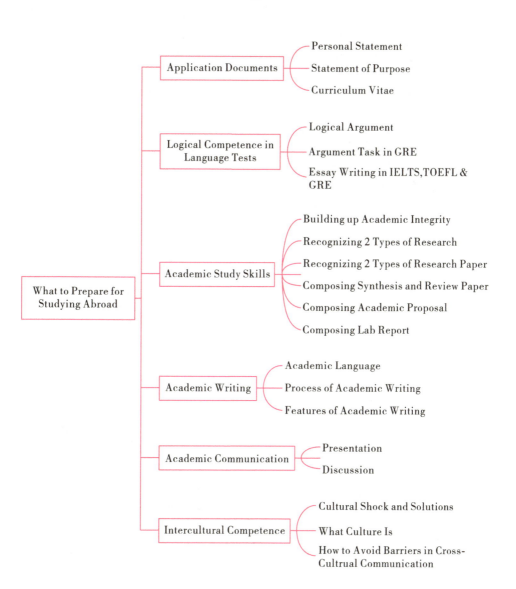

- What to Prepare for Studying Abroad
  - Application Documents
    - Personal Statement
    - Statement of Purpose
    - Curriculum Vitae
  - Logical Competence in Language Tests
    - Logical Argument
    - Argument Task in GRE
    - Essay Writing in IELTS,TOEFL & GRE
  - Academic Study Skills
    - Building up Academic Integrity
    - Recognizing 2 Types of Research
    - Recognizing 2 Types of Research Paper
    - Composing Synthesis and Review Paper
    - Composing Academic Proposal
    - Composing Lab Report
  - Academic Writing
    - Academic Language
    - Process of Academic Writing
    - Features of Academic Writing
  - Academic Communication
    - Presentation
    - Discussion
  - Intercultural Competence
    - Cultural Shock and Solutions
    - What Culture Is
    - How to Avoid Barriers in Cross-Cultrual Communication

Contents

Section 1

Applying for Universities

Unit 1

How to Write Personal Statement and Statement of Purpose

Learning objectives

- Knowing the differences between personal statement and statement of purpose
- Learning how to write personal statements with different purposes
- Learning how to write a statement of purpose

Task 1 Understanding what personal statement (PS) is

1.1 Visit the websites in Reference 1, 2 and 3 and note down the information related to the questions.

Questions:

1. What is a personal statement?
2. Who will be asked to write his/her personal statement?
3. What are the types of personal statement?
4. What are the tips for writing a personal statement?

Task 2 Learning how to write a personal statement

2.1 Read the personal statement below and figure out the use of each paragraph in writing.

Brent M. Ardaugh
Personal Statement
UCLA School of Public Health
Degree Objective: MPH

I would like to be a leader in the field of epidemiology. I have worked toward this goal by co-authoring an epidemiology course manual with my former professor, publishing articles for both professional and public audiences, and completing advance

Use of Para. 1 in writing:
· to highlight his future career goal

coursework in epidemiology, statistics, and biology. Some of my objectives for graduate school are to collaborate with UCLA faculty on research projects and to publish information in professional journals as well as public-oriented media types. Moreover, I would like to continue producing classroom resources for epidemiology faculty and students.

and objectives for his graduate school

Together with my former epidemiology professor, Dr. David R. Black, I co-authored a course manual that is currently being used by more than 50 undergraduate and graduate students at Purdue University – West Lafayette. This manual comprises over 650 PowerPoint slides and supplementary materials that introduce students to the basic concepts of epidemiology. Morbidity and mortality, screening tests, study designs, and causation are just a few of the topics discussed within the manual. Dr. Black and I intend to publish this manual for two reasons: to offer students a concise resource that they can use throughout their academic and professional careers and to provide faculty members with a complete "off-the-shelf" lecture series. I have attached sample excerpts from this manual to my application.

Use of Para. 2 in writing:

- to ____________

Writing is a population-based approach to preventing and controlling diseases. In the course of my studies, I published articles in the AMWA Journal, the official publication of the American Medical Writers Association (AMWA). Moreover, I wrote the medical writing resource for the Purdue University Online Writing Lab, a world-renowned instructional website for writers seeking consultation. One of my objectives for graduate school is to collaborate with the UCLA Office of Media Relations to publish articles on disease prevention and healthy living. In my articles, I would like to apply the knowledge from my UCLA epidemiology courses to provide readers with information that they can use to enhance their health and well-being. I would also enjoy opportunities to assist UCLA faculty in writing articles for professional journals and grant proposals for the National Institutes of Health (NIH). With my training in

Use of Para. 3 in writing:

- to ____________

medical journalism and NIH grant writing from the AMWA, I believe that I can better serve the Los Angeles community and support the advancement of epidemiology research at UCLA.

Epidemiology research enhances existing perception of people and their environments, and this perception leads to more effective methods to prevent and control diseases. I would like to research infectious disease epidemiology under Dr. Scott P. Layne, an epidemiology professor in your program. I prepared for this research by pursuing experience in microbiology, which consisted of lab work and studying in the classroom the physiology of these pathogens: Mycobacterium tuberculosis, Bacillus anthracis, Bordetella pertussis, Vibrio cholerae, HIV, Escherichia coli O157: H7, and Avian H5N1 Influenza. The above mentioned studying took place at the graduate level, where I was one of few undergraduate students. I will also be pursuing graduate level coursework in immunology, molecular biology, and genetics in spring 2009. I would like to apply the knowledge from these courses, in addition to the courses at UCLA, to develop strategies that protect populations from bioterrorism. Knowledge of how pathogens acquire virulent abilities, whether naturally or artificially induced, is necessary to ensure that preventative methods are adaptive to changes in pathogen evolution.

Use of Para. 4 in writing:

- to ____________

My second research interest is monitoring pathogens in human populations. Because of my curiosity, I was particularly impressed by a related research article by Dr. Layne entitled, "Human Influenza Surveillance: The Demand to Expand." This article was published in the *Emerging Infectious Diseases Journal* in 2006. In his article, Dr. Layne explained how a high throughput laboratory network can provide faster vaccine delivery methods and monitor changes in the epidemic influenza strains. I would look forward to working with Dr. Layne because it is an opportunity to expand my knowledge of medically relevant pathogens. This type of research has the potential to improve the effectiveness of the existing influenza protocol and to invest in

Use of Para. 5 in writing:

- to ____________

the future health and well-being of the world's population.

I am requesting admission to UCLA's MPH program in epidemiology. My previous epidemiology experience, academic preparation, and personal qualities have prepared me for the expectations of your program. My objective for graduate school is to combine rigorous academic study with hands-on experience, and I believe that Los Angeles and UCLA offer extraordinary opportunities for these endeavors. Lastly, I believe that I can contribute to your program through research, publishing, and multidisciplinary collaboration. My goal is to utilize the intellectual richness and diversity of UCLA to enhance the quality of life of the world's people.

Use of Para. 6 in writing:

- to ___________

2.2 Work in a small group of 3–4 members and discuss the questions below.

1. What will be included in a PS?
2. What else have your learned from this sample PS?
3. What should you prepare for your PS?

2.3 Read Reference 4 and answer the questions in it.

Task 3 Learn how to write a statement of purpose (SOP)

3.1 Visit the websites in Reference 5, 6 and 7, and note down the information related to the questions.

Questions:

1. What is a statement of purpose for?
2. Who is required to write this document?
3. What should be included in a statement of purpose?

3.2 Visit the website in Reference 8, select a university where you prefer having your graduate study and compose a statement of purpose.

Unit 2

How to Write Curriculum Vitae

Learning objectives

- Knowing the differences between a curriculum vitae and a resume
- Knowing the different types of curriculum vitae
- Learning how to write a curriculum vitae
- Learning language features of curriculum vitae

Task 1 Recognizing the differences between a curriculum vitae (CV) and a resume

1.1 Read the excerpts below and summarize the major differences between a CV and a resume by filling the form following the excerpts.

Excerpt 1

When to use a resume

In the United States, most employers use resumes for non-academic positions, which are one or two-page summaries of your experience, education, and skills. Employers rarely spend more than a few minutes reviewing a resume and successful resumes are concise with enough white space on the page to make it easy to scan.

For more information on developing your resume, please visit Optimal Resume and Cornell Career Services Career Guide. Students often find it helpful to review resumes from graduate students who got their first job outside of academe. To see example resumes, visit the PhD Career Finder in Versatile PhD.

When to use a curriculum vitae (CV)

A curriculum vitae (CV) is a longer synopsis of your educational and academic background as well as teaching and research experience, publications, awards, presentations, honors, and additional details. CVs are used when applying for academic, scientific, or research positions. International employers often use CVs as well.

(***Source***: https://gradschool.cornell.edu/academic-progress/pathways-to-success/prepare-for-your-career/take-action/resumes-and-cvs/)

Excerpt 2

How to prepare a resume for a master's program

The rules for writing a resume or curriculum vitae (CV) for graduate school are a little different than they are for the ordinary job hunt, but the end goal is still the same: you want to make it clear that your particular qualifications make you a good fit for this opportunity. These five tips will help you prepare the perfect resume for applying to a master's program:

1. Resume vs. curriculum vitae

Depending on what type of master's program you're applying to, you may be asked to prepare either a resume or a CV. For instance, Master of Arts in Teaching (MAT) or Master of Education (ME) programs can ask for either a resume or a CV, while MBA programs will ask for a resume. Both documents present a short history of your accomplishments, but a resume emphasizes professional accomplishments, while a CV emphasizes academic accomplishments. The two documents also tend to be formatted differently (for instance, a resume is usually one page while a CV is two pages), although neither document has a single standard format.

2. Emphasize education

Because you're applying for admission to an academic program, your resume or CV should emphasize academic accomplishments over professional accomplishments. You should also go into greater detail about your education than you would on your average professional resume. In addition to identifying your *alma mater*, degree earned and year of graduation, detail some of the more relevant classes you took and any academic honors you earned.

3. Include volunteer work and internships

If you're applying to graduate school straight out of undergrad, you may not have a lot of professional experience, but internships or volunteer work can tell admissions officers about your interests while demonstrating that you're willing to work hard for reasons other than immediate material rewards. If you're a professional looking to change fields, volunteer work or internships can also help to bridge the gap between your current profession and your desired field. Extracurricular activities are another way you can let admissions officers know about your personal passions and even your leadership experience.

4. Use language that makes an impact

A resume is no place to waste words. Think about the language you use, and choose words carefully to convey as much as you can about your accomplishments in the minimal amount of space a resume provides. The Internet is full of clichés about writing resumes, and you should be careful about which tips you use as you update your resume. For example, writing in the active voice is good because it is efficient and conveys a sense of energy, but buzzwords like "dynamic" and "detail-oriented" will only tip off admissions officers to the fact that you read an article about how to write a resume.

(***Source***: https://rossieronline.usc.edu/blog/how-to-prepare-a-resume-when-applying-to-a-masters-program/)

Excerpt 3

Resume vs. curriculum vitae: what's the difference?

Curriculum Vitae (CV) is Latin for "course of life." In contrast, resume is French for "summary." Both CVs & resumes:

- Are tailored for the specific job/company you are applying to
- Should represent you as the best qualified candidate
- Are used to get you an interview
- Do not usually include personal interests
- If you are applying for both academic as well as industry (private or public sector) positions, you will need to prepare both a resume and a CV.

Curriculum vitae vs. resume: format and content

The CV presents a full history of your academic credentials, so the length of the document is variable. In contrast, a resume presents a concise picture of your skills and qualifications for a specific position, so length tends to be shorter and dictated by years of experience (generally 1 – 2 pages).

CVs are used by individuals seeking fellowships, grants, postdoctoral positions, and teaching/research positions in postsecondary institutions or high-level research positions in industry. Graduate school applications typically request a CV, but in general are looking for a resume that includes any publications and descriptions of research projects.

In many European countries, CV is used to describe all job application documents, including a resume. In the United States and Canada, CV and resume are sometimes used interchangeably. If you are not sure which kind of document to submit, it is best to ask for clarification.

Resume	CV
· Emphasize skills · Used when applying for a position in industry, non-profit, and public sector · Is no longer than 2 pages, with an additional page for publications and/or poster presentations if highly relevant to the job · After 1 year of industry experience, lead with work experience and place education section at the or near the end, depending upon qualifications	· Emphasizes academic accomplishments · Used when applying for positions in academia, fellowships and grants · Length depends upon experience and include a complete list of publications, posters, and presentations · Always begins with education and can include name of advisor and dissertation title or summary... Also used for merit/tenure review and sabbatical leave

(***Source***: https://icc.ucdavis.edu/materials/resume/resumecv)

	CV	Resume
Origin of term	Comes form the Latin phrase *curriculum vitae* which literally means "________ ________"	Comes from the French word *résumé* which means "________________"
Purpose		
Content		
Length		

Task 2 Learning how to write different types of CV

2.1 Read Reference 9 (Sample CV 1) and Reference 10 (Sample CV 2) and figure out their type according to the information shown below.

Chronological CV	Skills-based CV
Information in each section is organized in reverse chronological order	Less emphasis on who he/she worked for and what job title was
To emphasize the consistency of experience · If you want to make it easier for potential readers · If you want to demonstrate growth and maturity throughout an organization · If you do not have many achievements	To highlight one's main skills, strength and expertise · If you do have these qualities · If you are mature enough and want to distract the attention from your age (for career application)

According to the information, I think
Sample CV 1 is a ________________ CV.
Sample CV 2 is a ________________ CV.

2.2 Read Reference 11 (Sample CV 3). Work in a small group of 3 – 4 members and discuss what type does this sample belong to and what strength this type of CV has. Take the notes while discussing.

➲ Your notes of the discussion:

2.3 Read the three sample CVs again and the website in Reference 12. Work in your group and discuss the question below.

Question: Generally speaking, what are the components of a CV?

Task 3 Learning the language features of CV

3.1 Read the excerpt of a CV below and then answer the question.

Allen Yan
(86)1338-1111-420
yhnasa@123. com

EDUCATION

...

EMPLOYMENT HISTORY

Dec. 2016 – Present, ITT Flygt Investment, China

Application Engineer, Sales & Marketing

· Achieve sales budget goals through application support and new industry market application research.

· Pay visits to end users and DI for seminars and technical presentations with salespersons or distributors while collecting marketing information and competitor information analysis.

July 2016 – Sept. 2016, Intel Products Co., Shanghai, China

CPU Assembly Engineer (Internship)

· Analyzed yield ratio trends, documented and solved current problems.

· Participated in and helped oversee the training of marketing, business process modeling, and analysis at Intel University.

· Developed and led a project review with multi-media animation, which was highly appreciated by department manager.

June 2015 – July 2015, GF Fund Management Co., LTD.

Campus Intern

· Analyzed investment principles and related financially derived products.

· Formulated the scheme of market popularization and network marketing.

AWARDS

...

Question:

According to the part "Employment History" in the CV above, what are the language features of a CV? You may highlight the verbs used to describe what the applicant has achieved before you answer the question. Pay attention to the tenses used.

3.2 Edit the problematic sentences below by correcting the inappropriate use of tenses.

- Have learned data collection methods: the focal and scan sampling methods, transects and behavioral observations.

  ____

  ____

- Developing a research proposal; have found appropriate methods; conducted research; analyzed and interpreted results; have written up a full research project and given an oral presentation.

  ____

  ____

- Assisted with reception duties, received out-patient prescriptions and delivered them to pharmacist for processing; informing and directing nursing staff when collecting medication or leaving in-patient prescriptions.

  ____

  ____

- Have befriended patients and relatives who needed extra support.

  ____

  ____

3.3 Rewrite the following description in parallel phrases by choosing the appropriate verbs provided. Make sure the tense is correct.

achieve	create	interview	obtain	devise	complete	prepare

"For my final-year project, I had to carry out a survey of patients' attitudes to health care services for the elderly. This involved interviewing 70 patients in hospital and in their homes. A database was used to keep track of data collected. This project was finished on time and was awarded a 2.1 grade."	➡ ________ and ________ a survey of patients' attitudes to health care services for the elderly as my final-year project ➡ ________ 70 elderly patients and ➡ ________ a substantial amount of data ➡ ________ a database to analyses and interpret this material ➡ ________ this project three weeks ahead of schedule and ________ a 2.1 grade

(***Source***: https://www.kent.ac.uk/careers/cv/actionverbs.htm)

Reflection

- CV writing, one type of academic writing, should be objective/depersonalized, concise and precise. When constructing a CV, you should omit ____________, use ____________ and keep the tense ____________ and accurate.
- I should prepare ____ for my CV if I plan to study abroad.

Reflection

[illegible]

Section 2

Preparing Logic for Language Tests

Unit 1

How to Do Argument Task in GRE

Learning objectives

- Learning what an argument is
- Learning how to analyze an argument
- Learning how to do the "Analyze an Argument" task in GRE

Task 1 Learning what an argument is

1.1 Read an excerpt from the book *Logical Reasoning* and understand the term "argument."

What is an argument?

The word argument has more than one meaning. In this book we will not use the word in the sense of being unpleasantly argumentative. Instead, it will mean at least one conclusion supported by one or more reasons, all of which are statements.

It takes only one person to have our kind of argument, not two. Saying that two people are "in an argument" means that there are two arguments, not one, in our sense of "argument." Each of the two persons has his or her own argument. In short, our word argument is a technical term with a more precise meaning than it has in ordinary conversation.

Statements that serve as reasons in an argument are also called premises. Nothing to do with the yard sign that says, "Keep off the premises." Any argument must have one or more premises. And it will have one or more "inference steps" taking you from the premises to the conclusion. The simplest arguments have just one step. Here is an example of a very simple argument that takes you to the conclusion in just one inference step from two premises:

If it's raining, we should take the umbrella.

It is raining.

So, we should take the umbrella.

1.2 Highlight the key ideas related to the term "argument" in the excerpt.

Task 2 Learning how to analyze an argument

2.1 Read the three examples/ways of analyzing an argument and figure out the elements in each way.

Example 1: CER (Claim, Evidence & Reasoning)

Claim: Air is matter. (finding/conclusion)

Evidence: We found that the weight of the ball increased each time we pumped more air into it. (result/data)

Reasoning: This shows that air has weight, one of the characteristics of matter. (theory; to connect evidence and claim)

Example 2: Syllogism (a standard form to rewrite an argument/to show the ways of reasoning)

P1 (major premise): Weight is one of the characteristics of matter.

P2 (minor premise): Air has weight (because we found...).

--(**Therefore**)

Conclusion: Air is matter.

Example 3: the Toulmin Model

Claim: Air is matter.

Evidence/Observation: We found that the weight of the ball increased each time we pumped more air into it.

Theory: Weight is one of the characteristics of matter.

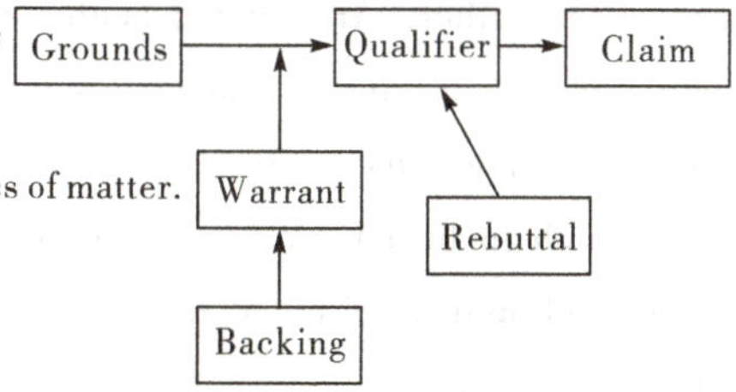

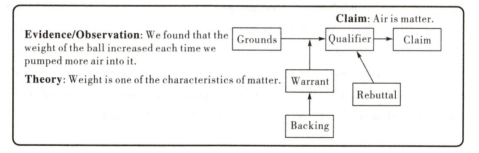

2.2 Read Reference 13 to learn the items as followed:

1. ways of logical reasoning
2. validity of logical reasoning
3. some common logical fallacies

2.3 Search and study the term "the Toulmin Model" on the Internet.

2.4 Work in a group of 3 or 4 members and share what you have learned from 2.2 and 2.3.

2.5 Read the passage below and analyze the argument in it by using CER, ways of reasoning and the Toulmin Model.

Some mothers suffer agony from incessantly crying babies during the first three months of life. Nothing the parents do seems to stem the flood. They usually conclude that there is something radically, physically wrong with the infants and try to treat them accordingly. They are right, of course, that there is something physically wrong; but it is probably effect rather than cause. The vital clue comes with the fact that this so-called "colic" crying ceases, as if by magic, around the third or fourth month of life. It vanishes at just the point where the baby is beginning to be able to identify its mother as a known individual.

A comparison of the parental behavior of mothers with crying babies and those with quieter infants gives the answer. The former is tentative, nervous and anxious in their dealing with their offspring. The latter are deliberate, calm and serene. The point is that even at this tender age, the baby is acutely aware of differences in tactile "security" and "safety", on the one hand, and tactile "insecurity" and "alarm" on the other. An agitated mother cannot avoid signaling her agitation to her new-born infant. It signals back to her in the appropriate manner, demanding protection from the cause of the agitation. This only serves to increase the mother's distress, which in turn increases the baby's crying. Eventually the wretched infant cries itself sick and its physical pains are then added to the sum total of its already considerable misery.

All that is necessary to break the vicious circle is for the mother to accept the situation and become calm herself. Even if she cannot manage this (and it is almost impossible to fool a baby on this score) the problem corrects itself, as I said, in the third or fourth month of life, because at that stage the baby becomes imprinted on the mother, and instinctively begins to respond to her as the "protector". She is no longer a disembodied series of agitating stimuli, but a familiar face. If she continues to give agitating stimuli, they are no longer so alarming because they are coming from a known source with a friendly identity. The baby's growing bond with its parents then calms the mother and automatically reduces her anxiety. The "colic" disappears.

(***Source***: Desmond Morris, *The Naked Ape*, *New York*: *Dell Publishing Co. Inc.*, 1967, pp. 98-9)

Task 3 Learning how to do the "Analyze an Argument" task in GRE

3.1 Read the introduction to this writing task adapted from the official site of GRE and try to explain the terms which are shaded.

The "Analyze an Argument" task assesses your ability to understand, analyze and evaluate arguments and to clearly convey your evaluation in writing. You are presented with a brief passage in which the author makes a case for some course of action or interpretation of events by presenting claims backed by reasons and evidence. Your task is to discuss the logical soundness of the author's case according to the specific instructions by critically examining the line of reasoning and the use of evidence. The specific instructions could be one of the followings:

· Write a response in which you discuss what specific evidence is needed to evaluate the argument and explain how the evidence would weaken or strengthen the argument.

· Write a response in which you examine the stated and/or unstated assumptions of the argument. Be sure to explain how the argument depends on these assumptions and what the implications are if the assumptions prove unwarranted.

· Write a response in which you discuss what questions would need to be answered in order to decide whether the recommendation and the argument on which it is based are reasonable. Be sure to explain how the answers to these questions would help to evaluate the recommendation.

· Write a response in which you discuss what questions would need to be answered in order to decide whether the advice and the argument on which it is based are reasonable. Be sure to explain how the answers to these questions would help to evaluate the advice.

· Write a response in which you discuss what questions would need to be answered in order to decide whether the recommendation is likely to have the predicted result. Be sure to explain how the answers to these questions would help to evaluate the recommendation.

· Write a response in which you discuss what questions would need to be answered in order to decide whether the prediction and the argument on which it is based are reasonable. Be sure to explain how the answers to these questions would help to evaluate the prediction.

· Write a response in which you discuss one or more alternative explanations that could rival the proposed explanation and explain how your explanation(s) can plausibly account for the facts presented in the argument.

(***Source***: https://www.ets.org/gre/revised_general/about/content/analytical_writing)

3.2 Read the excerpt from the section "Preparing for the Argument Task" on the official website of GRE and figure out what academic skills are required by this writing task.

Since the Argument task is meant to assess analytical writing and informal reasoning skills that you have developed throughout your education, it has been designed neither to require any specific course of study nor to advantage students with a particular type of training.

Many college textbooks on rhetoric and composition have sections on informal logic and critical thinking that might prove helpful, but even these might be more detailed and technical than the task requires. You will not be expected to know specific methods of analysis or technical terms.

For instance, in one topic an elementary school principal might conclude that new playground equipment has improved student attendance because absentee rates have declined since it was installed. You will not need to see that the principal has committed the post hoc, ergo propter hoc fallacy; you will simply need to see that there are other possible explanations for the improved attendance, to offer some common-sense examples and to suggest what would be necessary to verify the conclusion. For instance, absentee rates might have decreased because the climate was mild. This would have to be ruled out in order for the principal's conclusion to be valid.

Although you do not need to know special analytical techniques and terminology, you should be familiar with the directions for the Argument Task and with certain key concepts, including the following:

· alternative explanation—a competing version of what might have caused the events in question that undercuts or qualifies the original explanation because it too can account for the observed facts

· analysis—the process of breaking something (e.g., an argument) down into its component parts in order to understand how they work together to make up the whole

· argument—a claim or a set of claims with reasons and evidence offered as support; a line of reasoning meant to demonstrate the truth or falsehood of something

· assumption—a belief, often unstated or unexamined, that someone must hold in order to maintain a particular position; something that is taken for granted but that must be true in order for the conclusion to be true

· conclusion—the end point reached by a line of reasoning, valid if the reasoning is sound; the resulting assertion

· counterexample—an example, real or hypothetical, that refutes or disproves a statement in the argument

· evaluation—an assessment of the quality of evidence and reasons in an argument and of the overall merit of an argument

An excellent way to prepare for the "Analyze an Argument" task is to practice writing on some of the published argument topics. There is no one way to practice that is best for everyone. Some prefer to start practicing without adhering to the 30-minute time limit. If you follow this approach, take all the time you need to evaluate the argument. Regardless of the approach you take, consider the following steps:

· Carefully read the argument and the specific instructions—you might want to read them more than once.

· Identify as many of the argument's claims, conclusions and underlying assumptions as possible and evaluate their quality.

· Think of as many alternative explanations and counterexamples as you can.

· Think of what specific additional evidence might weaken or lend support to the claims.

· Ask yourself what changes in the argument would make the reasoning more sound.

Write down each of these thoughts. When you've gone as far as you can with your evaluation, look over the notes and put them in a good order for discussion (perhaps by numbering them). Then write an evaluation according to the specific instructions by fully developing each point that is relevant to those instructions. Even if you choose not to write a full essay response, you should find it helpful to practice evaluating a few of the arguments and sketching out your responses.

When you become quicker and more confident, you should practice writing some Argument responses within the 30-minute time limit so that you will have a good sense of how to pace yourself in the actual test. For example, you will not want to discuss one point so exhaustively or to provide so many equivalent examples that you run out of time to make your other main points.

You might want to get feedback on your response(s) from a writing instructor, philosophy teacher or someone who emphasizes critical thinking in his or her course. It can also be informative to trade papers on the same topic with fellow students and discuss each other's responses in terms of the scoring guide. Focus not so much on the "right scores" as on seeing how the responses meet or miss the performance standards for each score point and what you need to do to improve.

(***Source***: https://www.ets.org/gre/revised_general/prepare/analytical_writing/argument/prepare_task)

3.3 Read the two cases below and analyze them with the instructions below.

Instructions:

- Carefully read them and figure out the line of reasoning, deductive or inductive.
- Identify as many of the argument's claims, conclusions and underlying assumptions as possible and evaluate their quality.
- Think of as many alternative explanations and counterexamples as you can.
- Think of what specific additional evidence might weaken or lend support to the claims.
- Ask yourself what changes in the argument would make the reasoning sounder or more cogent.

Case 1

Woven baskets characterized by a particular distinctive pattern have previously been found only in the immediate vicinity of the prehistoric village of Palea and therefore were believed to have been made only by the Palean people. Recently, however, archaeologists discovered such a "Palean" basket in Lithos, an ancient village across the Brim River from Palea. The Brim River is very deep and broad, and so the ancient Paleans could have crossed it only by boat, and no Palean boats have been found. Thus it follows that the so-called Palean baskets were not uniquely Palean.

Case 2

The following appeared in a letter to the editor of the Balmer Island Gazette.

"On Balmer Island, where mopeds serve as a popular form of transportation, the population increases to 100,000 during the summer months. To reduce the number of accidents involving mopeds and pedestrians, the town council of Balmer Island should limit the number of mopeds rented by the island's moped rental companies from 50 per day to 25 per day during the summer season. By limiting the number of rentals, the town council will attain the 50 percent annual reduction in moped accidents that was achieved last year on the neighboring island of Seaville, when Seaville's town council enforced similar limits on moped rentals."

3.4 Analyze the requirements below and figure out what should be written in each part in your response to each argument task in 3.3.

Case 1: Write a response in which you discuss what specific evidence is needed to evaluate the argument and explain how the evidence would weaken or strengthen the argument.

Case 2: Write a response in which you discuss what questions would need to be answered in order to decide whether the recommendation is likely to have the predicted result. Be sure to explain how the answers to these questions would help to evaluate the recommendation.

Questions:

1. What should be written in the introduction? Why?
2. How many paragraphs are there in the body part? Why?
3. How is the writing concluded? Why?

3.5 Finish each argument task writing within 30 minutes.

3.6 Read "Scoring Guide for the Argument Task in GRE" below, and highlight the different diction in it.

Score 6: Outstanding	Score 5: Strong
In addressing the specific task directions, a 6 response presents a cogent, well-articulated examination of the argument and conveys meaning skillfully. A typical response in this category: · clearly identifies aspects of the argument relevant to the assigned task and examines them insightfully · develops ideas cogently, organizes them logically and connects them with clear transitions · provides compelling and thorough support for its main points · conveys ideas fluently and precisely, using effective vocabulary and sentence variety · demonstrates superior facility with the conventions of standard written English (i. e., grammar, usage and mechanics), but may have minor errors	In addressing the specific task directions, a 5 response presents a generally thoughtful, well-developed examination of the argument and conveys meaning clearly. A typical response in this category: · clearly identifies aspects of the argument relevant to the assigned task and examines them in a generally perceptive way · develops ideas clearly, organizes them logically and connects them with appropriate transitions · offers generally thoughtful and thorough support for its main points · conveys ideas clearly and well, using appropriate vocabulary and sentence variety · demonstrates facility with the conventions of standard written English, but may have minor errors

3.7 Here is a topic for argument analysis writing. Read it.

In surveys Mason City residents rank water sports (swimming, boating and fishing) among their favorite recreational activities. The Mason River flowing through the city is rarely used for these pursuits, however, and the city park department devotes little of its budget to maintaining riverside recreational facilities. For years there have been complaints from residents about the quality of the river's water and the river's smell. In response, the state has recently announced plans to clean up Mason River. Use of the river for water sports is therefore sure to increase. The city government should for that reason devote more money in this year's budget to riverside recreational facilities.

Write a response in which you examine the stated and/or unstated assumptions of the argument. Be sure to explain how the argument depends on the assumptions and what the implications are if the assumptions prove unwarranted.

3.8 Work in a small group of 3 – 4 members and discuss the questions: Do you think the argument in 3.7 is strong or weak? Why?

3.9 According to the scoring guide given, grade the two sample responses to the argument you just have read and discussed.

Note: All responses are reproduced exactly as written, including errors, misspellings, etc., if any.

Essay Response 1—Your score ________

The author of this proposal to increase the budget for Mason City riverside recreational facilities offers an interesting argument but to move forward on the proposal would definitely require more information and thought. While the correlations stated are logical and probable, there may be hidden factors that prevent the City from diverting resources to this project.

For example, consider the survey rankings among Mason City residents. The thought is that such high regard for water sports will translate into usage. But, survey responses can hardly be used as indicators of actual behaviors. Many surveys conducted after the winter holidays reveal people who list exercise and weight loss as a top priority. Yet every profession does not equal a new gym membership. Even the wording of the survey results remain ambiguous and vague. While water sports may be among the residents' favorite activities, this allows for many other favorites. What remains unknown is the priorities of the general public. Do they favor these water sports above a softball field or soccer field? Are they willing to sacrifice the municipal golf course for better riverside facilities? Indeed the survey hardly provides enough information to discern future use of improved facilities.

Closely linked to the surveys is the bold assumption that a cleaner river will result in increased usage. While it is not illogical to expect some increase, at what level will people begin to use the river? The answer to this question requires a survey to find out the reasons our residents use or do not use the river. Is river water quality the primary limiting factor to usage or the lack of docks and piers? Are people more interested in water sports than the recreational activities that they are already engaged in? These questions will help the city government forecast how much river usage will increase and to assign a proportional increase to the budget.

Likewise, the author is optimistic regarding the state promise to clean the river. We need to hear the source of the voices and consider any ulterior motives. Is this a campaign year and the plans a campaign promise from the state representative? What is the timeline for the clean-up effort? Will the state fully fund this project? We can imagine the misuse of funds in renovating the riverside facilities only to watch the new buildings fall into dilapidation while the state drags the river clean-up.

Last, the author does not consider where these additional funds will be diverted from. The current budget situation must be assessed to determine if this increase can be afforded. In a sense, the City may not be willing to draw money away from other key projects from road improvements to schools and education. The author naively assumes that the money can simply appear without forethought on where it will come from.

Examining all the various angles and factors involved with improving riverside recreational facilities, the argument does not justify increasing the budget. While the proposal does highlight a possibility, more information is required to warrant any action.

Essay Response 2—Your score ________

While it may be true that the Mason City government ought to devote more money to riverside recreational facilities, this author's argument does not make a cogent case for increased resources based on river use. It is easy to understand why city residents would want a cleaner river, but this argument is rife with holes and assumptions, and thus, not strong enough to lead to increased funding.

Citing surveys of city residents, the author reports city resident's love of water sports. It is not clear, however, the scope and validity of that survey. For example, the survey could have asked residents if they prefer using the river for water sports or would like to see a hydroelectric dam built, which may have swayed residents toward river sports. The sample may not have been representative of city residents, asking only those residents who live upon the river. The survey may have been 10 pages long, with 2 questions dedicated to river sports. We just do not know. Unless the survey is fully representative, valid, and reliable, it can not be used to effectively back the author's argument.

Additionally, the author implies that residents do not use the river for swimming, boating, and fishing, despite their professed interest, because the water is polluted and smelly. While a polluted, smelly river would likely cut down on river sports, a concrete connection between the resident's lack of river use and the river's current state is not effectively made. Though there have been complaints, we do not know if there have been numerous complaints from a wide range of people, or perhaps from one or two individuals who made numerous complaints. To strengthen his/her argument, the author would benefit from implementing a normed survey asking a wide range of residents why they do not currently use the river.

Building upon the implication that residents do not use the river due to the quality of the river's water and the smell, the author suggests that a river clean up will result in increased river usage. If the river's water quality and smell result from problems which can be cleaned, this may be true. For example, if the decreased water quality and aroma is caused by pollution by factories along the river, this conceivably could be remedied. But if the quality and aroma results from the natural mineral deposits in the water or surrounding rock, this may not be true. There are some bodies of water which emit a strong smell of sulphur due to the geography of the area. This is not something likely to be affected by a clean-up. Consequently, a river clean up may have no impact upon river usage. Regardless of whether the river's quality is able to be improved or not, the author does not effectively show a connection between water quality and river usage.

A clean, beautiful, safe river often adds to a city's property values, leads to increased tourism and revenue from those who come to take advantage of the river, and a better overall quality of life for residents. For these reasons, city government may decide to invest in improving riverside recreational facilities. However, this author's argument is not likely significantly persuade the city government to allocate increased funding.

3.10 Work in pairs or a small group of 3 members, share your grading results and evaluate your work in 3.5 with the scoring guide offered by the official website in Reference 14.

Unit 2

How to Write Essay in IELTS, TOEFL & GRE

Learning objectives

- Learning how to analyze the arguments in the essay writing tasks in IELTS, TOEFL & GRE
- Learning to express your argument

Task 1 Learning how to analyze the arguments in the essay writing tasks in IELTS, TOEFL &GRE

1.1 Read the directions of the essay writing tasks in IELTS, TOEFL and GRE and analyze the arguments by using CER, logical reasoning and the Toulmin Model.

IELTS:

In Task 2, test takers write an essay in response to a point of view, argument or problem.

You should spend about 40 minutes on this task.

Write about the following topic:

The first car appeared on British roads in 1888. By the year 2000 there may be as many as 29 million vehicles on British roads. *Alternative forms of transport should be encouraged and international laws introduced to control car ownership and use.* To what extent do you agree or disagree?

Give reasons for your answer and include any relevant examples from your knowledge or experience.

Write at least 250 words.

(*Source*: https://www.ielts.org/about-the-test/sample-test-questions#tab-4)

TOEFL:

For the second task, you will demonstrate your ability to write an essay in response to a question that asks you to express and support your opinion about a topic or issue. In an actual test, your essay would be scored on the quality of your writing. This includes the development of your ideas, the organization of your essay, and the quality and accuracy of the language you use to express your ideas.

You would have 30 minutes to plan, write, and revise your essay. Test takers with disabilities may request a time extension. Typically, an effective response contains a minimum of 300 words.

Question: *Do you agree or disagree with the following statement?* *A teacher's ability to relate well with students is more important than excellent knowledge of the subject being taught.* Use specific reasons and examples to support your answer.

(*Source*: https://www.ets.org/Media/Tests/TOEFL/pdf/SampleQuestions.pdf)

GRE: the Issue Task

The "Analyze an Issue" task assesses your ability to think critically about a topic of general interest and to clearly express your thoughts about it in writing. Each Issue topic makes a claim that can be discussed from various perspectives and applied to many different situations or conditions. Your task is to present a compelling case for your own position on the issue. Before beginning your written response, be sure to read the issue and the instructions that follow the Issue statement. Think about the issue from several points of view, considering the complexity of ideas associated with those views. Then, make notes about the position you want to develop and list the main reasons and examples you could use to support that position.

Following is a sample Issue task that you might see on the test:

As people rely more and more on technology to solve problems, the ability of humans to think for themselves will surely deteriorate. Discuss the extent to which you agree or disagree with the statement and explain your reasoning for the position you take. In developing and supporting your position, you should consider ways in which the statement might or might not hold true and explain how these considerations shape your position.

The GRE raters scoring your response are not looking for a "right" answer—in fact, as far as they are concerned, there is no correct position to take. Instead, the raters are evaluating the skill with which you address the specific instructions and articulate and develop an argument to support your evaluation of the issue.

(*Source*: https://www.ets.org/gre/revised_general/prepare/analytical_writing/issue/)

Task 2 Analyzing samples of essay writing

2.1 Read the sample response to Question 2 in TOEFL writing section and answer the questions.

QUESTION 2, RESPONSE A, SCORE OF 5

1. I remember every teacher that has taught me since I was in Kindergarten. If a friend wants to know who our first grade teacher was in elementary school, all they have to do is ask me. The teachers all looked very kind and understanding in my eyes as a child. They had special relationships with nearly each and every one of the students and were very nice to everyone. That's the reason I remember all of them.

2. A teacher's primary goal is to teach students the best they can about the things that are in our textbooks and more important, how to show respect for one another. They teach us how to live a better life by getting along with everyone. In order to do that, the teachers themselves have to be able to relate well with students.

3. My parents are teachers too. One teaches Plant Biology and one teaches English, but that's not the reason I'm calling them "teachers." They are teachers because they teach me how to act in special situations and how to cooperate with others. I have a brother, and my parents use different approaches when teaching us. They might scold my brother for surfing the internet too long because he doesn't have much self-control and they need to restrain him. He almost never studies on his own

and is always either drawing, playing computer games, or reading. On the other hand, they never tell me off for using the computer too long. I do my own work when I want and need to because that brings me the best results and my parents understand that. They know that I need leisure time of my own and that I'll only play until needed. My parents' ability to relate well with my brother and I allows them to teach, not just the subject they teach but also their excellent knowledge on life.

4. Knowledge of the subject being taught is something taken for granted, but at the same time, secondary. One must go through and pass a series of courses and tests in order to become a teacher. Any teacher is able to have excellent knowledge of their subject but not all teachers can have the ability to relate well with students.

5. A teacher's primary goal is to teach students the best they can about how to show respect for one another, so teachers use different approaches when teaching, and knowledge of the subject being taught is secondary. For these reasons, I claim with confidence that excellent knowledge of the subject being taught is secondary to the teacher's ability to relate well with their students.

(***Source***: https://www.ets.org/s/toefl/pdf/writing_practice_sets.pdf)

Questions:

1. What is the writer's opinion on the topic or issue "A teacher's ability to relate well with students is more important than excellent knowledge of the subject being taught"?

2. How does the writer respond to the essay question "Do you agree or disagree with the following statement?"

3. What way of development is adopted in this writing?

   A. From general to specific

   B. From specific to general

4. What are the specific reasons and examples are given by the writer?

5. What way of reasoning is used in Para. 4? Present the way of reasoning by using the Standard Form.

6. Can you find the language mistakes in this writing? What are they? How can we get a full/satisfactory score?

2.2 Read the sample response to Task 2 in IELTS writing section and answer the questions.

The transport has been one of the most important problem for the last two centuries. The problem began with the development and the growing of the cities.

Before the eigth century the people lived in small villages or towns and did not have necessity To To too far. The people did not worry about the time to arrive in some where.

Nowadays the situation changed. Many cars on the streets and many people need to go to any place. The numbers of car has increased and as a result there are many problems: pollution, noise, car accident, unsuffinent car park and petroleum problem.

On the other hand, people use car to go anywhere: to work, to travel, to spent holiday and to amusement. Meanwhile the car is important the cities must have another solution. It is important to organise its using and to meet alternative ways.

In big cities there are some alternatives like undergrounds (metro), coach, train and bycicles. In China and Cuba for example they use a lot of bycicles for substituting the cars or coaches. /---

It would be better to think about others differents kinds of transport. In Brasil the Government has talked about transport on the rivers. In this country there are many rivers where it is possible to go to different places. In general they are flat rivers.

Another kind of transport is car that uses solar energy. Probably they don't have pollution problem and it is cheaper than others car.

In conclusion, the transport is a social problem in big cities but its solution depend on new. Technologies, others kind of energy and political aspects.

Text of the sample response

The transport has been one of the most important problem for The last Two centuries. The problem began with the development and the growing of the cities.

Before the eighth century the people lived in small villages or tows and did not have necessity to go too far, the people did not worry about the time to arrive in home where.

Nowadays the situation changed. Many cars on the streets and many people need to go to any place. The numbers of car has increased and as a result there are many problems: pollution, noise, car accident, unsufficient car park and petroleum problem.

On the other hand, people use car to go anywhere: to work, to travel, to spent holiday and to amusement. Meanwhile the car is important the cities must have another solution. It is important to organize its using and to meet alternative ways.

In by cities there are some alternatives like undergrounds (metro), coach, train and bycicles. In China and Cuba for example they use a lot of bycicles for substituting the cars and coaches.

It would be better to think about others different kinds of transport. In Brazil the government has talked about transport on the rivers. In this country there are many rivers where it is possible to go to different places. In general, they are flat rivers.

Another kind of transport is car that uses solar energy. Probably they don't have pollution problem and it is cheaper than others car.

In conclusion, The transport is a social problem in big cities but its solution depend on new technologies, others kind of energy and political aspects.

(***Source***: https://www.ielts.org/-/media/pdfs/113313_ac_sample_scripts.ashx?la=en)

Questions:

1. How does the writer respond the essay question "To what extent do you agree or disagree?"
2. What is the writer's position?
3. What way of development is adopted in this writing?
   A. From general to specific
   B. From specific to general
4. What are the reasons given by the writer to support his/her position?
5. How do you improve this writing with the reference to the comments offered by the test official website?

Band 6 (Full score: 9)

There are quite a lot of ideas and while some of these are supported better than others, there is an overall coherence to the answer. The introduction is perhaps slightly long and more time could have been devoted to answering the question. The answer is fairly easy to follow and there is good punctuation. Organizational devices are evident although some areas of the answer become unclear and would benefit from more accurate use of connectives. There are some errors in the structures but there is also evidence of the production of complex sentence forms. Grammatical errors interfere slightly with comprehension.

(***Source***: https://www.ielts.org/-/media/pdfs/113313_ac_sample_scripts.ashx? la = en)

6. Which type of claim is the statement "Alternative forms of transport should be encouraged and international laws introduced to control car ownership and use"?
   A. Claim of fact
   B. Claim of value
   C. Claim of policy
7. According to this writing, how can this type of claim be proved in writing?

2.3 Read the sample response to the Issue Task in GRE writing section and answer the questions.

Essay Response—Score 6 (Full score: 6)

1. The statement linking technology negatively with free thinking plays on recent human experience over the past century. Surely there has been no time in history where the lived lives of people have changed more dramatically. A quick reflection on a typical day reveals how technology has revolutionized the world. Most people commute to work in an automobile that runs on an internal combustion engine. During the workday, chances are high that the employee will interact with a computer that processes information on silicon bridges that are 0.09 microns wide. Upon leaving home, family members will be reached through wireless networks that utilize satellites orbiting the earth. Each of these common occurrences could have been inconceivable at the turn of the 19th century.

2. The statement attempts to bridge these dramatic changes to a reduction in the

ability for humans to think for themselves. The assumption is that an increased reliance on technology negates the need for people to think creatively to solve previous quandaries. Looking back at the introduction, one could argue that without a car, computer, or mobile phone, the hypothetical worker would need to find alternate methods of transport, information processing and communication. Technology short circuits this thinking by making the problems obsolete.

3. However, this reliance on technology does not necessarily preclude the creativity that marks the human species. The prior examples reveal that technology allows for convenience. The car, computer and phone all release additional time for people to live more efficiently. This efficiency does not preclude the need for humans to think for themselves. In fact, technology frees humanity to not only tackle new problems, but may itself create new issues that did not exist without technology. For example, the proliferation of automobiles has introduced a need for fuel conservation on a global scale. With increasing energy demands from emerging markets, global warming becomes a concern inconceivable to the horse-and-buggy generation. Likewise dependence on oil has created nation-states that are not dependent on taxation, allowing ruling parties to oppress minority groups such as women. Solutions to these complex problems require the unfettered imaginations of maverick scientists and politicians.

4. In contrast to the statement, we can even see how technology frees the human imagination. Consider how the digital revolution and the advent of the internet has allowed for an unprecedented exchange of ideas. WebMD, a popular internet portal for medical information, permits patients to self research symptoms for a more informed doctor visit. This exercise opens pathways of thinking that were previously closed off to the medical layman. With increased interdisciplinary interactions, inspiration can arrive from the most surprising corners. Jeffrey Sachs, one of the architects of the UN Millenium Development Goals, based his ideas on emergency care triage techniques. The unlikely marriage of economics and medicine has healed tense, hyperinflation environments from South America to Eastern Europe.

5. This last example provides the most hope in how technology actually provides hope to the future of humanity. By increasing our reliance on technology, impossible goals can now be achieved. Consider how the late 20th century witnessed the complete elimination of smallpox. This disease had ravaged the human race since prehistorical days, and yet with the technology of vaccines, free thinking humans

dared to imagine a world free of smallpox. Using technology, battle plans were drawn out, and smallpox was systematically targeted and eradicated.

6. Technology will always mark the human experience, from the discovery of fire to the implementation of nanotechnology. Given the history of the human race, there will be no limit to the number of problems, both new and old, for us to tackle. There is no need to retreat to a Luddite attitude to new things, but rather embrace a hopeful posture to the possibilities that technology provides for new avenues of human imagination.

Questions:

1. How does the writer respond the essay question "To what extent do you agree or disagree?"
2. What is the writer's position?
3. What is the writer's claim?
4. What way of development is adopted in this writing?
   A. From general to specific
   B. From specific to general
5. Do you think the writer use the same way to prove his claim/position as the ways used in the samples of TOEFL and IELTS writing tasks? Why?
6. Analyze the argument in this essay with the Toulmin Model and ways of reasoning.

Reflection

What should be considered when you are planning your essay writing in IELTS, TOEFL or GRE?

Task 3 Learning to express your argument

Analyze the statement in the directions below with the Toulmin Model and ways of reasoning and figure out the outline for your essay writing.

In any field of endeavor, it is impossible to make a significant contribution without first being strongly influenced by past achievements within that field.

Write a response in which you discuss the extent to which you agree or disagree with the statement and explain your reasoning for the position you take. In developing and supporting your position, you should consider ways in which the statement might or might not hold true and explain how these considerations shape your position.

Reflection

What have you learned from this section about essay writing?

Section 3

Preparing for Academic Study

Unit 1

Building Academic Integrity

Learning objectives

- Knowing what academic integrity is
- Building up your academic integrity
- Knowing different ways to cite sources

Task 1 Building your academic integrity

1.1 Visit the websites in Reference 15, 16, 17 and 18. Read them carefully and answer the questions.

Questions:

What is academic integrity?

What is it important for study and research?

Why do you think there is a problem of academic dishonesty in our country?

Do you have any solutions to this problem? What are they?

What are YOU going to do to foster YOUR academic integrity from now on?

1.2 Visit the website in Reference 19, read it carefully, note down the important information you think and answer the question below.

Question:

According to what is mentioned on this web page, how do you understand the word "original" in the title?

1.3 Visit the website in Reference 20, read every item carefully, note down the information which can help you be academically honesty and answer the question.

Question:

What are the academic skills, including academic language skills required to hold your academic integrity?

1.4 Watch Video 1 and note down the information related to the question.

Questions:

Why do we cite?

Task 2 Recognizing different formats for citing information

2.1 Revisit the website in Reference 20 and review the section about "Citing Sources".

2.2 Read three papers in Reference 21, 22 and 23 and figure out each paper's features of citation and reference. The three papers are

- Reference 21: Engineering Research Paper (Paper 1)
- Reference 22: Dough and Gluten (Paper 2)
- Reference 23: Disrupting White Normativity (Paper 3)

Papers	Citation	Reference
Paper 1		
Paper 2		
Paper 3		

Task 3 Writing a review paper on the topic "academic integrity"

☺ Definitely, you can use more sources about this topic to finish your writing job.

☺ Before writing, learn next section in this book.

Unit 2

How to Read a Paper

Learning objective

- Learning how to read a paper efficiently

Task 1 Learning how to read papers efficiently

1.1 Read Reference 24—"How to Read a Paper" and figure out the use of each part in this writing.

Part	Use in writing
Abstract	To do a concise summary of the whole paper
Introduction	
The three-pass approach	
Doing a literature survey	
Related work	
Acknowledgments	
References	

1.2 Read this paper again, finish a key-word outline in the table below and write a summary. The length of the summary is 1/3 or 1/4 of the original text.

Outline
Title: Research question: Introduction: Body: 1.   2.   3.   Conclusion:

Task 2 Practicing efficient reading strategies

2.1 Read Reference 21 and 22 by using the first two passes and note down the important information you get from each of the two passes.

Unit 3

Recognizing Types of Research and Paper

Learning objectives

- Recognizing different types of research
- Recognizing different types of research paper

Task 1 Recognizing types of research

1.1 Carefully read the charts below and explain what each chart informs us about doing research.

Chart 1

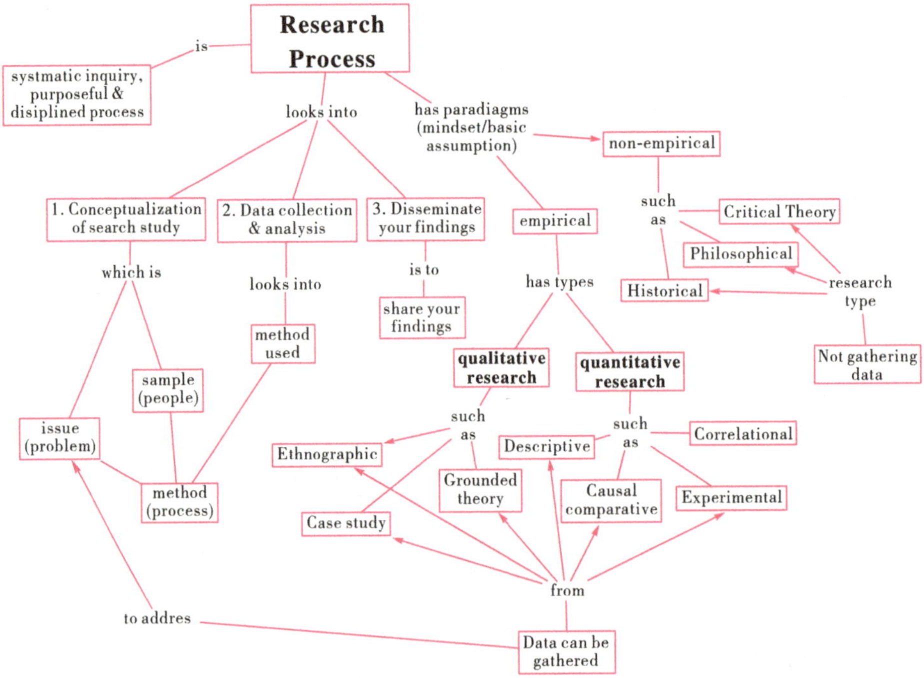

Chart 2

Primary Research Methods & Techniques

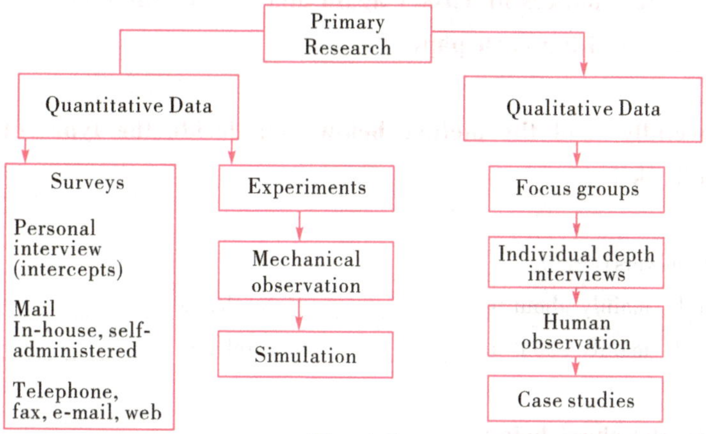

Chart 3

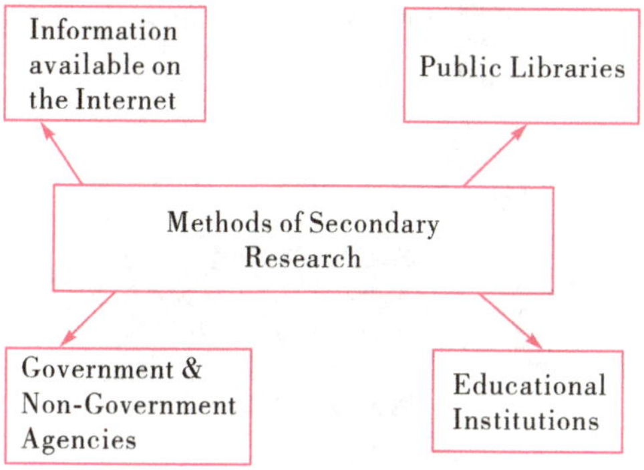

1.2 Read Reference 21 and 22 again and draw a flow chart of the process of each research.

1.3 Answer the questions according to the two papers in 1.2.

Q1: According to these two papers, what research methods are used in the two studies?

Q2: What is the research question of each study? What is studied?

Q3: What are the purposes of doing the two studies?

Q4: How did the researchers collect the data for their study?

Q5: Do you think the two studies are of the same type of research? Why?

1.4 Work in a group of 3 or 4 members and do a group presentation to show your answers to 1.3.

Task 2 Recognizing types of paper

2.1 Read the 2 papers in Task 1 again and answer the question: What is the purpose of writing each paper?

2.2 Carefully read the picture below and decide the types of the two papers in Task 1.

Your answers:

Paper 1, mainly about a ____________ research, is ____________ paper.

Paper 2, mainly about a ____________ research, is ____________ paper.

2.3 Read the chart below.

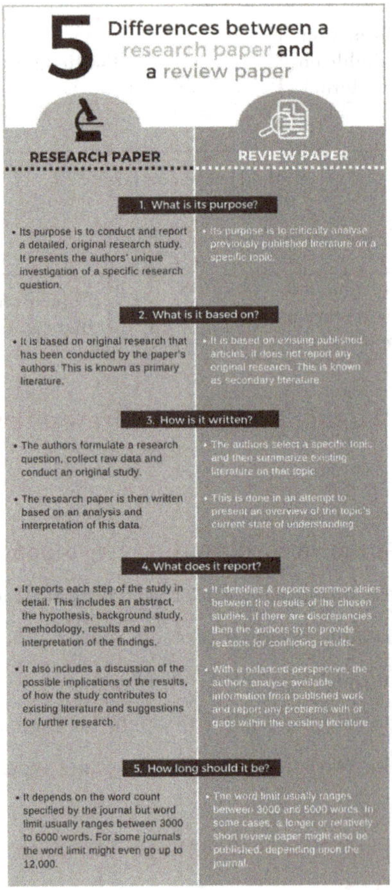

Unit 4

Recognizing Types of Review Paper

Learning objectives
- Recognizing different types of review papers
- Recognizing different structures of review papers

Tip: Please DON'T read the sources of the papers below until Task 1 is totally accomplished!!! Your teacher will announce the right time to read the sources.

Task 1 Identifying different types of review

1.1 Read the abstracts of the three review papers and fill in the table below.

	Paper 1	**Paper 2**	**Paper 3**
Research topic			
Research question			
Purpose(s)/Objective(s) of research			
Findings of research			
Meanings of research			

Paper 1

Community-based organizations in the health sector

Michael G Wilson, John N Lavis and Adrian Guta

Abstract

Community-based organizations are important health system stakeholders as they provide numerous, often highly valued programs and services to the members of their community. However, community-based organizations are described using diverse terminology and concepts from across a range of disciplines. To better understand the literature related to community-based organizations in the health sector (i. e., those working in health systems or more broadly to address population or public health issues), we conducted a [...] review by using an iterative process to identify existing literature, conceptually map it, and identify gaps and areas for future inquiry. We searched 18 databases and conducted citation searches using 15 articles to identify relevant literature. All search results were reviewed in duplicate and were included if they addressed the key characteristics of community-based organizations or networks of community-based organizations. We then coded all included articles based on the country focus, type of literature, source of literature, academic discipline, disease sector, terminology used to describe organizations and topics discussed.

We identified 186 articles addressing topics related to the key characteristics of community-based organizations and/or networks of community-based organizations. The literature is largely focused on high-income countries and on mental health and addictions, HIV/AIDS or general/unspecified populations. A large number of different terms have been used in the literature to describe community-based organizations and the literature addresses a range of topics about them (mandate, structure, revenue sources and type and skills or skill mix of staff), the involvement of community members in organizations, how organizations contribute to community organizing and development and how they function in networks with each other and with government (e. g., in policy networks). Given the range of terms used to describe community-based organizations, this scoping review can be used to further map their meanings/definitions to develop a more comprehensive typology and understanding of community-based organizations. This information can be used in further investigations about the ways in which community-based organizations can be engaged in health system decision-making and the mechanisms available for facilitating or supporting their engagement.

(*Source*: http://www.health-policy-systems.com/content/10/1/36)

Paper 2

The value and impact of information provided through library services for patient care

Weightman AL, Williamson J; Library & Knowledge Development Network (LKDN) Quality and Statistics Group

Abstract

OBJECTIVE: An updated review was carried out of research studies looking at the value and impact of library services on health outcomes for patients and time saved by health professionals.

METHODS: A comprehensive systematic search was undertaken of the published literature to September 2003 in ERIC, LISA, MEDLINE, PREMEDLINE, EMBASE, the Cochrane Controlled Trials Register and Google. Some handsearching was carried out, reference lists were scanned and experts in the field were contacted. Twenty-eight research studies of professionally led libraries for health-care staff, including clinical librarian projects, met the inclusion criterion of at least one health or "time saved" outcome. Papers were critically appraised using internationally accepted criteria. Data were extracted and results were summarized using a narrative format as the studies were heterogeneous and precluded a statistical analysis.

RESULTS: There is evidence of impact from both traditional and clinical librarian services. The higher quality studies of traditional services measured impacts of 37% - 97% on general patient care, 10% - 31% on diagnosis, 20% - 51% on choice of tests, 27% - 45% on choice of therapy and 10% - 19% on reduced length of stay. Four studies of clinical librarian projects suggested that professionals saved time as a result of clinical librarian input, and two of these studies showed evidence of cost-effectiveness. However, the clinical librarian studies were generally smaller, with poorer quality standards.

CONCLUSIONS: Research studies suggest that professionally led library services have an impact on health outcomes for patients and may lead to time savings for health-care professionals. The available studies vary greatly in quality but the better quality studies also suggest positive impacts. Good practice can be gathered from these studies to guide the development of a pragmatic survey for library services that includes the direct effects for patients among the outcome measures.

(***Source***: onlinelibrary. wiley. com/doi/full/10. 1111/j. 1471-1842. 2005. 00549. x)

Paper 3

Past, present and future trends of non-radiative wireless power transfer

S. Y. Ron Hui

Abstract

Although non-radiative wireless power transfer (WPT) was invented over a century ago, it has regained research and development interests in 1980s. Over the last 15 years, WPT has appeared as an "emerging" technology that has attracted wide-spread attention in both academia and industry. Because of the long history of WPT research and developments, researchers of the modern days often do not know some historical milestones of WPT. This paper aims at providing a brief history of some key concepts and technologies that pave the way for modern WPT research and applications. A few misconceptions of WPT technologies are particularly highlighted so that new researchers entering this new research field can avoid such pitfalls. Finally, some discussions on present and future trends of WPT are included.

(***Source***: https://ieeexplore. ieee. org/document/7911098)

1.2 Check your answers with your partners.

1.3 Read the three quotations about types of review as followed, and then decide the type of each of the three review papers in 1.1.

"A systematic review attempts to identify, appraise and synthesize all the empirical evidence that meets pre-specified eligibility criteria to answer a given research question. Researchers conducting systematic reviews use explicit methods aimed at minimizing bias, in order to produce more reliable findings that can be used to inform decision making."

(***Source***: *Cochrane Handbook for Systematic Reviews of Interventions*)

"... scoping reviews [aim to] map key concepts underpinning a research area and the main sources and types of evidence available, and [are rapidly] undertaken as stand-alone projects in their own right, especially where an area is complex or has not been reviewed comprehensively before... scoping reviews are often conducted to examine previous research activity, disseminate findings, identify gaps in the research and/or determine the value of conducting a full systematic review..."—Wilson et al, 2012

(***Source***: http://hlwiki.slais.ubc.ca/index.php/Scoping_reviews)

"A narrative review summarizes different primary studies from which conclusions may be drawn into a holistic interpretation contributed by the reviewers' own experience, existing theories and models. Results are of a qualitative rather than a quantitative meaning... Narrative reviews should make the search criteria and the criteria for inclusion explicit. It critically evaluates the specific topic of research."

(***Source***: *A Guide for Writing Scholarly Articles or Reviews for the Educational Research Review*)

Your answers:

Paper 1 is a ____________________ review.

Paper 2 is a ____________________ review.

Paper 3 is a ____________________ review.

1.4 Think about what you have done in 1.1 and 1.3 and have a group discussion on the question: Which type of review paper is the paper "A Review on Rheological Properties and Measurements of Dough and Gluten"?

Task 2 Learning to structure review papers

2.1 Read the three papers in Reference 25, 26 and 27, and note down the subtitles in the following chart.

Paper 1	Paper 2	Paper 3

2.2 Think about the answers in 2.1 and have a discussion about the structure of the three review papers.

☺ You may answer in the way given below.

Generally speaking, the three review papers are composed of...

However, there are some differences in their structure... As to me, the reason(s) is/are...

☺ You can take notes while you are discussing.

2.3 Your review paper on "academic integrity" can be written in any type of review paper. Work in a group of 3 – 4 members and discuss the purposes and structures of each type.

Narrative Review	Scoping Review	Systematic Review

Unit 5

How to Write a Literature Review

Learning objective

- Learning how to read literature
- Learning how to evaluate the sources of information
- Learning how to synthesize the information from the different sources
- Recognizing different types of literature review

Task 1 Knowing how to write a literature review

1.1 Read Reference 28—"Active Reading Strategies". Highlight the important information while reading.

1.2 Read the website in Reference 29 by using the active reading strategies offered by Reference 28 in 1.1.

1.3 Work in a group of 3 or 4 members and discuss what you have learned from the website. Take notes when necessary in the discussion.

1.4 Individually answer the questions below according to what you have learned from the website or your own knowledge.

1. What is literature?
2. What is a literature review?
3. Why should we write a literature review?
4. What are the key steps of writing a literature review?
5. What should we do when searching for relevant literature?
6. How do we evaluate the sources?
7. What should we take notes and cite the sources when evaluating and selecting sources?
8. What are the connections and relationships between the sources you've read?
9. What are the approaches to organizing the body of a literature review?

10. What is the structure of a literature review? Generally, how many parts are there in each literature review?
11. How many kinds of literature review are there?
12. What is a good literature review?
13. How do we analyze sources?
14. How do we synthesize sources?
15. How do we critically evaluate sources?

1.5 Watch Video 2 and notes down the information about how to evaluate the source?

1.6 Read the website in Reference 30 and notes down the information about how to evaluate the digital sources?

Task 2 Learning synthesis

2.1 Read the websites in Reference 31 and 32, and take notes about "synthesis".

2.2 Use a synthesis matrix to collect the notes about the two websites in 2.1.

2.3 Figure out the relationships between the two sources.

2.4 Outline the structure of a synthesis related to the two sources. The questions here may help you finish this job.

1. Which type of synthesis is your work? Explanatory syntheses or argumentative syntheses?
2. What is the use of the three parts in your writing: introduction, body and conclusion?
3. Which approach will be used to outline the body part?
4. How many paragraphs are there in the body part? What is the topic sentence of each paragraph? What is the rationale of your paragraphing and topic sentences' writing?

Task 3 Understanding the skills required in TOEFL Writing Task One and learning how to write it

3.1 Read the requirements of Writing Task One: Integrated Writing in TOFEL.

- Directions: Give yourself 3 minutes to read the passage.
- Directions: Listen to a lecture.
- Directions: Give yourself 20 minutes to plan and write your response. Your response is judged on the quality of the writing and on how well it presents the points in the lecture and their relationship to the reading passage. Typically, an effective response will be 150 to 225 words. You may view the reading passage while you respond.
- Response time: 20 minutes
- **Question: Summarize the points made in the lecture, being sure to explain how they cast doubt on specific points made in the reading passage.**

3.2 Analyze the question and figure out what academic skills are required when accomplishing this writing tasks and what the structure of the writing is.

3.3 Share your thoughts in your group of 3 – 4 members.

3.4 Try your ways in practicing writing this task.

- **Directions**: Give yourself 3 minutes to read the passage.

Scan physical evidence remains of the first human domestication of grain. Still, there is enough to conclude that ancient peoples, motivated by the nutritional value of bread or cakes made of wild wheat, looked for controlled ways to grow it to provide a consistent food supply. Three related discoveries are likely to have led to the introduction of bread as the first grain-based food.

The first discovery was that wheat could be prepared for use by grinding. People probably began consuming wheat by chewing it raw. Because wheat is very hard, they gradually discovered that it was less trouble to eat if crushed to paste between two stones. The result would have been the ancestor of the drier, more powdery wheat flour we use today.

> From there, it was a short step to the next breakthrough—baking the simplest bread, which requires no technology but fire. Loaves of wheat paste, when baked into bread, could be stored for long periods, certainly longer than raw seeds. This kept the food value of weak available for an extended period after it had been harvested.
>
> Finally, ancient peoples found that, if the paste was allowed to sit in the open, yeast spores from the air settled on it and began fermenting the wheat. This natural process of fermentation caused bubbles to form in the wheat paste, suggesting that it would be lighter in texture and even easier to eat when baked.

- **Directions:** Listen to a lecture. (Audio 1)
- **Response time:** 20 minutes
- **Question:** Summarize the points made in the lecture, being sure to explain how they cast doubt on specific points made in the reading passage.

3.5 Read Appendix Ⅰ and check your writing.

Checklist

- Purpose of writing
- Relationship between the passage and the lecture
- Structure of writing

  Para. 1: To introduce ______________________

  Para. 2: To ____________ ______________________

  Para. 3: To ____________ ______________________

  Para. 4: To ____________ ______________________

  Q: Is the conclusion paragraph needed? Why?
- Language features
  - Tenses
  - Reporting language
  - Paraphrasing

Unit 6

How to Cite Information

Learning objectives

- Learning what APA Format is
- Learning how to cite information in your paper writing
- Recognizing different types of citation
- Learning how to cite the information from different sources
- Learning the use of tenses when citing information
- Learning the uses of reporting verbs

Task 1 Learning APA Format

1.1 Globally understand APA Format by skimming its official website in Reference 33.

1.2 Work out the rules used in in-text citation in the paper below.

It is widely known that elephants fear cheese, and will flee at the first whiff of it ("Elephants stampede", 2003). What is not yet well understood is why this phenomenon occurs. For more than a decade academics have been researching this perplexing topic. Their work constitutes part of the booming new discipline known as pachydermo-fromagology, which is defined as "the study of elephant-cheese interactions" (Concise Oxford dictionary, 2004). This paper will evaluate existing research and theories, and argue that none of them satisfactorily explain the data which has been gathered so far.

Source F

Source D

That elephants fear cheese was an accidental discovery made by the noted elephantologist G. Coleman (Coleman, 1984). The story of the discovery is now famous, but worth repeating:

Source B

After a hard morning following the herd, I had just sat down under a tree for lunch and unwrapped a particularly delectable chunk of cheddar sent up from the base camp. Suddenly I heard an enormous trampling sound, and when I looked up, the entire herd was gone. (Coleman, 1988, p. 160)

Source C, page 160

His discovery, while dismissed at the time, was subsequently corroborated by other researchers. Several studies (Gibson & Sturgess, 1987; Gibson, Sturgess, & Bates, 1989) have confirmed the phenomenon, and that it occurs among both African and Asian elephants. A recent report by the Elephant Research Institute (2001) established that smell is the primary means elephants detect cheese, and that they will ignore large pieces of cheese if tightly wrapped. Meanwhile a French cheese expert asserts on his website that elephants do not flee from French cheese, only the lesser cheeses of other nations. "Zee creatures, zey have good taste, non?" he writes (Gouda, n. d., para. 2).

Sources G, H

Source E

Source I, Introduction, para. 2

Recently, a new theory has exploded on the scene and caused quite a stink. Based on several clever experiments, K. Maas (Maas, 2003, p. 468) has claimed that in fact elephants do not fear cheese at all, but instead fear the mice which are attracted to cheese. However, this theory, which she calls the Maas Mouse Hypothesis (MMH), has not yet been widely accepted. One researcher (Sturgess, 2004a; 2004b) has published a series of articles roundly denouncing the MMH, and the debate has even spilled over into the popular press (Achison, 2004).

Source J, page 468

Source K, L

Source A

What are we to make of this controversy?

Reference

Achison, C. L. (2004, April). A ripe and weighty issue: An interview with Monica Sturgess. *Cheese Lovers World*, *6* (4), 12 – 13.

A

Coleman, G. J. (1984). An odd behaviour observed among the species Elephas maximus. *Journal of Trunked Mammal Studies*, *23*, 421-429. B

Coleman, G. J. (1988). *Underfoot: Ten years among the elephants*. New York: Oxford University Press. C

Concise Oxford dictionary, 11th ed. (2004). Oxford University Press. Retrieved October 20, 2004, from Oxford Reference Online database. D

Elephant Research Institute, Simon Fraser University. (2001) *Smell versus sight: detection of cheese by elephants*. Retrieved November 1, 2004, from http://www.sfu.ca/eri/reports/00107elephants.pdf E

Elephants stampede, 7 cheese-lovers trampled. (2003, November 22). Vancouver Sun, p. A1, A8. F

Gibson, C. N. & Sturgess, M. N. (1987). Elephant fleeing behaviour confirmed. *Journal of Elephantology*, *16*, 239-245. Retrieved October 27, 2004, from Academic Search Elite database. G

Gibson, C. N. Sturgess, M. N., & Bates, A. T. (1989). Experiments with cheese effects on Elephas maximus and Elephas africanus. *Journal of Elephantology*, *18*, 120-134. Retrieved October 27, 2004, from Academic Search Elite database. H

Gouda, A. N. (n. d.) *Commentary of a report about cheese and les elephants*. Retrieved October 23, 2004, from http://www.mondedefromage.fr/elephants.html I

Maas, K. A. (2003). The missing link: elephants, mice, and J cheese. *International Journal of Rodentia Research*, *56*, 459-471. Retrieved October 31, 2004, from http://www.elsevierpublisher.com/ijrr/56/4/maas.htm

Sturgess, M. N. (2004a). Of mice and cheese (Part 1). *Journal of Trunked Mammal Studies*, *43*, 10-15. K

Sturgess, M. N. (2004b). Of mice and cheese (Part 2). *Journal of Trunked Mammal Studies*, *43*, 219-225. L

1.3 Think about the questions below and have a discussion with your partner.

Q1: How many kinds of in-text citation are there?

Q2: Why are the different kinds of in-text citation used in academic writing? Explain your answers.

1.4 Read the paragraph below and do the exercise following it.

Market-based organization

The discipline and practice of marketing has undergone a period of transformation over the last four decades (Varadarajan, 1999). For instance, Morgan (1996) suggested that during the 1960s the marketing concept was thought to be the savior of companies. The 1970s witnessed a challenge to this because marketing did not respond to greater issues in society. During the 1980s marketing caused discontent by over-segmenting markets and overstating the value of consumers' expressed needs. McKenna (1991:49) states that in the 1990s, marketing could be considered as "everything". Since then, many developments have taken place but none has been timelier than Sheth and Sisodia's (1999) assertion that academicians must revisit marketing's law-like generalizations. As a result of this, marketing function has now evolved into a component of the whole organisation. For instance, it now considers more global issues such as the Mission and Vision of the organisation (Vorhies & Yarbrough, 1998).

This paragraph is adapted from:

Morgan, R. E., McGuiness T. & Thorpe E. R. (2000). The contribution of marketing to business strategy formation: a perspective on business performance gains, *Journal of Strategic Marketing 8/4*: 341 – 362

➡ Suppose someone cites Morgan in his/her writing and uses the **integral citations** or **narrative citation**, which may show the author's name in a variety of ways, as shown in these examples:

1. According to Morgan (1996), during the 1960s the marketing concept was thought to be the savior of companies.

2. Morgan (1996) suggests that during the 1960s the marketing concept was thought to be the savior of companies.

3. Morgan's theory (1996) suggests that during the 1960s the marketing concept was thought to be the savior of companies.

4. The role of the marketing concept as the savior of companies was suggested by Morgan (1996).

Match each of the patterns below with the appropriate sentence above.

a. Researcher as subject of the sentence Sentence ________

b. Researcher as part of a reporting phrase/clause Sentence ________

c. Researcher as agent of a passive sentence Sentence ________

d. Researcher as part of a possessive noun phrase Sentence ________

1.5 Read the following sections from four original articles that are cited in the paragraph in 1.4 and decide whether the citations in that paragraph have been plagiarized. If so, decide why.

Sections from four original articles	Have the citations in the paragraph in 1.4 been plagiarized? If yes, why?	Rewrite the citations identified as being plagiarized.
Varadarajan (1999) The theory and activity of marketing has changed significantly over the past forty years.		

(To be continued on next page)

Morgan (**1996**) During the 1960s the marketing concept was thought to be the savior of companies. The 1970s witnessed a challenge to this because marketing did not respond to greater issues in society. Furthermore, during the 1980s marketing caused discontent by over-segmenting markets and overstating the value of consumers' expressed needs.		
Sheth and Sisodia's (**1999**) Marketing's law-like generalizations must be revisited by academicians.		
Vorhies and Yarbrough (**1998**) The role of the marketing function has changed significantly from its segmented position in the large, bureaucratic hierarchy of organizations in the 1950s and 1960s. Instead it has been incorporated into the structure of current organizations. For example, the Mission and Vision of the organization are now taken into consideration.		

Task 2 Learning to quote information

2.1 Visit the website in Reference 34 and learn when to use quotation.

2.2 Consider the following sentences taken from a student essay about Henry David Thoreau's "Civil Disobedience". In each one, a reporting verb is missing. Review the list of reporting verbs below and choose one to introduce each of the following quotations from Thoreau's essay. Is Thoreau arguing any of these statements, reporting, stating a belief or concluding? Make sure the verb you choose reflects the stance Thoreau takes in each particular statement.

Suggested Verbs			
acknowledge	claim	emphasize	note
add	comment	grant	observe
agree	confirm	illustrate	point out
assert	dispute	imply	reason

A. As Henry David Thoreau ____________________, "I think that we should be men first, and subjects afterward."

B. "I cannot for an instant recognize that political organization as my government which is the slave's government also," the abolitionist Thoreau ____________________.

C. Morally opposed to both the Mexican-American war and slavery, Thoreau ____________________ that "[u]nder a government which imprisons unjustly, the true place for a just man is also a prison."

D. "Cast your whole vote, not a strip of paper merely, but your whole influence," Thoreau ____________________ to the American people. "A minority is powerless while it conforms to the majority; it is not even a minority then; but it is irresistible when it clogs by its whole weight."

E. "I have paid no poll tax for six years," Thoreau ____________________.

2.3 Work in your group of 3 – 4 members and discuss your answers to 2.2.

Task 3 Learning more about in-text citation

3.1 Reread the paper "A Review on Rheological Properties and Measurements of Dough and Gluten" in Reference 30 and further understand the different uses of the two types of in-text citation.

3.2 Carefully read the definitions of the reporting verbs used in citing sources and classify them according to the tones they imply.

- Admit: Suggests that the author only reluctantly offers the information you're discussing
- Agree: Indicates that the author is echoing a viewpoint already established in your paper
- Argue: Indicates that the author is taking a stand on an issue

- Assume: Looks at an author's underlying reasons for making a particular claim; if applicable, you might follow a quote introduced by "assume" with an explanation of why this assumption is accurate or inaccurate
- Believe: Introduces source material that seems more opinionate than argument-based
- Blame: Indicates that the author is assigning responsibility for a negative outcome
- Censure: Introduces source material that include a reprimand of some sort
- Characterize: Suggests that the author includes a description that has an agenda (in other words, the author describes something in a particular way to make the reader come to a particular conclusion)
- Claim: Like "argue," indicates that the author is taking a stand on an issue
- Classify: Introduces a passage that breaks a topic down into categories
- Commend: Indicates that the author is praising someone or is assigning responsibility for a positive outcome
- Conclude: Suggests that the author reaches a decision after studying the topic carefully
- Condemn: Introduces a passage that attacks or denounces someone
- Consider: Suggests that the author entertained other perspectives or counterarguments
- Criticize: Indicates that the author is pointing out flaws or weaknesses in another person's arguments or actions
- Decide: Suggests that the author reaches a conclusion after studying the topic carefully
- Define: Indicates that the author is assigning a specific meaning to a given term or idea
- Deny: Indicates that the author is refuting another author's arguments, data, etc.
- Depict: Like "characterize," suggests that the author includes a description that has an agenda (in other words, the author describes something in a particular way to make the reader come to a particular conclusion)
- Describe: Indicates that the author is offering a verbal picture of a setting or a situation
- Determine: Like "conclude" and "decide" suggests that the author has reached a conclusion after careful study
- Discover: Indicates that the author has found out something new
- Doubt: Suggests that the author is uncertain about a particular idea or argument
- Evaluate: Suggests that the author has made a judgment of worth regarding a topic

- Explain: Implies that the author carefully breaks down the topic
- Hypothesize: Indicates that the author has proposed an original argument
- Imply: Indicates that an author is only suggesting a particular point, rather than coming right out and saying it
- Indicate: Like "say" or "state," shows that the author is putting forth information (NEUTRAL)
- Infer: Indicates that the author has reached a conclusion by drawing out logical information about a topic
- Maintain: Like "claim," suggests that the author is taking a stand on an issue
- Present: Indicates that the author is putting forth information
- Presume: Like "assume," looks at an author's underlying reasons for making a particular claim; if applicable, you might follow a quote introduced by "presume" with an explanation of why this presumption is accurate or inaccurate
- Prove: Indicates that an author definitively supported his/her argument
- Reveal: Suggests that the author is unveiling new information
- Show: Like "explain," implies that the author carefully breaks down the topic
- State: Indicates that the author is putting forth information (NEUTRAL)
- View: Like "believe," introduces source material that seems more opinionate than argument-based

(***Source***: www. temple. edu/wc 2)

3.3 Reread the paper and figure out the uses of reporting verbs.

3.4 Reread the paper and learn what tense (s) are used when sources are cited.

3.5 Check your answers in your group of 3 – 4 members.

Task 4 Put it together

Finish your review paper about "academic integrity" by using the skills you have learned in Units 2 – 6 in this section.

Unit 7

How to Write an Academic Proposal

Learning objectives

- Understanding what academic proposal is
- Learning how to create a strong research question
- Learn how to write an academic proposal

Task 1 Understanding what academic proposal is

1.1 Search the websites in Reference 35, 36, 37 and 38, and note down the relevant information in the table below with key words.

- Web 1 (Reference 36):
- Web 2 (Reference 37):
- Web 3 (Reference 38):
- Web 4 (Reference 39):

	Role of proposal in doing research	Purpose of writing	Tips of writing	Elements of a proposal
Web 1				
Web 2				

(To be continued on next page)

(continued)

	Role of proposal in doing research	Purpose of writing	Tips of writing	Elements of a proposal
Web 3				
Web 4				

1.2 Share your collection of information in 1.1 in your group of 3 – 4 members.

Task 2 Learning how to create a strong research question

2.1 Answer the following questions:

Q1: Do you know how to create a strong research question?

Q2: If yes, how do you do it?

Q3: If no, what questions do you have to help you to develop this academic skill?

2.2 Search the websites in Reference 39, 40, 41, and 42, and use a matrix to collect the information related to "research question".

2.3 Present your matrix in your group of 3 – 4 members and give your comments on the matric made by your group members after the presentation.

2.4 Watch video 3 and figure out some research questions in it.

Task 3 Learn how to write an academic proposal

3.1 Visit the websites in Reference 43 and 44, and learn the components of a proposal.

3.2 Read Proposal Sample 1 in Reference 45 and Proposal Sample 2 in Reference 46 and answer the questions below.

Q1: Who are the writers of the two proposals?

Q2: Who are the readers of the two proposals?

Q3: How many parts are there in each sample? What are they?

Q4: What are the differences between the two samples?

3.3 Highlight the "phraseological 'nuts and bolts' of writing" in the two samples. You may refer to Task 4, Unit 1, Section 4 to understand the term "phraseological 'nuts and bolts' of writing".

Task 4 Putting it together

4.1 Write a proposal about the research on the cultural analysis of the movies mentioned in Task 3, Unit 1, Section 5. This is an individual job.

4.2 Write a proposal about your research on academic integrity required in Task 3, Unit 1, Section 3. This is an individual job.

Unit 8

How to Write a Lab Report

Learning objectives

- Learning how to structure a lab report
- Learning how to write different parts in a lab report

Task 1 Learning how to structure a lab report

1.1 Read the abstract of a lab report which addresses the research question "What size electric heating element is installed in a given water heater?" Identify the use of abstract in writing and what can be included in an abstract by answering the questions below.

Abstract

In our experiment, we solved for the work put into the water through electric heating element. An electric water heater was attached to a domestic water supply. Water flowed through the device, and its temperature was measured at entrance and exit at volumetric flow rates ranging from 500 mL/min to 3000 mL/min. Linear regression was used to find the average work applied to the system, 920 ± 130W. The results had a relatively large error, $\approx 15\%$, which may be attributed to the assumption of an adiabatic system. To account for this error, future work would be to determine the heat loss to the environment, which we neglected, using an energy balance.

Questions	Phraseological "nuts and bolts" of writing
1. What specific problem/research question was being addressed? Why was the experiment done?	In our experiment, we solved for the work...
2. How was the experiment carried out? What methods were used to solve the problem/answer the question?	
3. What has been found in this experiment? What results were obtained?	

(To be continued on next page)

(continued)

Questions	Phraseological "nuts and bolts" of writing
4. What do the results mean? So what? How do they answer the overall question or improve our understanding of the problem?	

Question:

What tenses are used in this section? Why?

1.2 Read the section of Introduction of the lab report mentioned in 1.1 and find out what it explains by matching the numbered sentence(s) with the options given below.

Introduction

1 A fundamental concept in mechanical engineering is the first law of thermodynamics, which states that internal enemy is conserved in a control volume. 2 This law has many applications in engineering, such as: heat exchangers, pumps, turbines, HVAC mixing and refrigeration cycles. It is used to understand the states of fluids as they enter or leave a control volume. 3 The general form of the first law is Equation 1.

$$\frac{dE}{dt} = \dot{Q} - \dot{W} + \sum_{in} \dot{m}_{in}\left(h_{in} + \frac{V_{in}^2}{2} + gz_{in}\right) - \sum_{out} \dot{m}_{out}\left(h_{out} + \frac{V_{out}^2}{2} + gz_{out}\right)$$

Equation 1: First Law

4 This energy balance states that heat transfer into a system ($\dot{Q}$) less the work out of a system ($\dot{W}$) plus the mass flow out ($\dot{m}_e$) times its internal energy less the mass flow in ($\dot{m}_i$) times its internal energy is equal to the energy storage term ($\frac{dE}{dt}$).

5 In this experiment, the first law of thermodynamics will be used with an electric water heater to answer the research question, 6 "What size electric element is used in a given water electric heater?" We hypothesize it will be close to 1 kW, which is the approximate average of most other water heaters of similar size and shape.

S1	A. report question
S2	B. purpose
S3	C. explanation of the equation
S4	D. brief explanation of the relevant theory
S5	E. background / theory
S6	F. governing equation

Question:

What tenses are used in this section? Why?

1.3 Study the annotated section of Methods of the lab report in 1.1 and summarize what should be included in this section.

Methods

Tap water was passed through the electric water heater at flow ranging from 50 ml/min to 300 ml/min. the input and output temperatures were recorded. The information was then used with the first law to approximate the work by the heating element performed on the system. (***Experiment Overview instantly informs reader.***)

The experimental apparatus includes a plastic container, the heater, measuring 6" × 6" × 12". One half inch ports were located at the upstream and downstream sides of the heater and labelled 'in' and 'out'. An electric power cord is attached to the top side of the heater and supplies 120Vac to the heater inside. The input to the heater was connected to a domestic water supply, which provided 46.1 ± 0.4 ℉ water once at steady state. The output from the water heater was run through a Dwyer flowmeter, which measured and controlled flowrate, and then exited into a sink. Temperature readings were made at the inlet and exit ports of the heater with T thermocouples and an Omega temperature indicator. The experimental setup is shown in Figure 1. (***Apparatus description with the sketch enables reader to visualize the experiment.***)

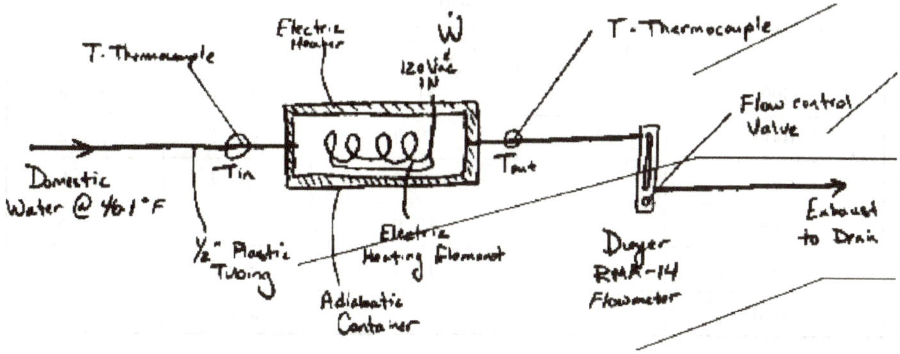

Figure 1: Apparatus

A detailed list of the equipment used in this experiment and their uncertainties are shown in Table 1.

Table 1: Equipment

Equipment	Uncertainty
Electric Heater	n/a
Dwyer RMA – 14 Rotameter	+/– 50 ml/min
Omega T Thermocouples	+/– 1.8 °F
Omega Tem Readout	n/a

The tap water entering the system was varied between 500 ml/min and 3000 ml/min, in 500 ml/min increments. At each flowrate, the inlet and outlet temperature were measured ten times when the system appeared to reach steady state. (***Step-by-step procedure***)

Question:

What tenses are used in this section? Why?

1.4 Study the section of Results of the lab report in 1.1 and learn what this section demonstrates by discussing the questions.

Results

The water heated up as it passed through the heat exchanger and the temperature change appeared to correlate inversely with the flow rate. The exit temperature of the water appear in Table 2. (***Q: What does this section begin with***?) Inlet water temperature reached steady state at 46.1 ±0.4 °F.

Table 2

Flowrate [ml/min]	T_{out} [℉]	$T_{out}-T_{in}$ [℉]	ΔT_{out} [℉]	$\Delta T_{out,\ Tot}$ [℉]
500	87.2	41.1	±0.1	±1.8
1000	75.4	29.3	±0.2	±1.8
1500	67.0	20.9	±0.4	±1.8
2000	63.1	17.0	±0.3	±1.8
2500	59.3	13.2	±1.2	±2.2
3000	56.1	8.7	±2.2	±2.8

(***Q*****: *Why is the table presented here*?)**

The uncertainty, ΔT_{out}, is the random precision error, using n = 10 samples, 9 d. f., and $t_{0.025,9}$ = 2.262 form the t distribution. Equation 2 statistically approximates the true mean, μ, with the samples collected.

$$\mu = \bar{x} \pm t_{0.025,9}\frac{s_x}{\sqrt{n}} \quad (95\ \text{C. L.})$$

Equation 2

In the above equation, S_x is the standard deviation of the sample collected and $\bar{x}$ is the mean value of the samples. The total uncertainty in outlet temperature, $\Delta T_{out,\ Tot}$ was calculated by combining both the random uncertainty of the temperature readings and the bias uncertainty of the thermocouples given by the manufacture of +/−1.8 ℉ using root sum of squares (RSS). This process is shown in Equation 3.

$$\Delta T_{out,\ Tot} = \sqrt{\Delta T^2_{out,random} + \Delta T^2_{out,bias}}$$

Equation 3

(***Q*****: *What is the relationship between the equations and the sentences before and after them*?)**

The result and uncertainties were then graphed to visualize the data. Error bars were created using for temperature using data from Table 2 and errors for volumetric flowrate were from the manufacturer, shown in Table 1. This data is shown in Figure 2.

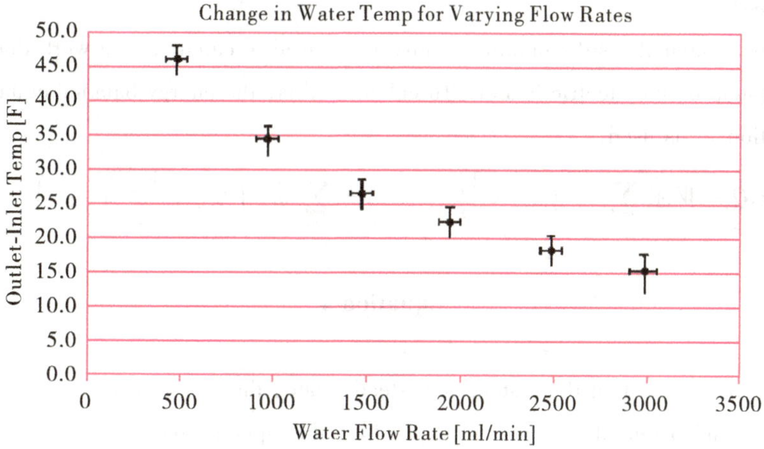

Figure 2 Experimental Results

The trend in the data seems logical as the faster the flowrate, the smaller the temperature difference between inlet and outlet will be. Uncertainty in temperature measurement grew as the flowrate increased, perhaps due to turbulent flow at the heater exit or a temperature profile in the water resulting from incomplete mixture.

(***Q: Do you think whether it is needed to describe what the data means in this section***?)

Question:

What tenses are used in this section? Why?

1.5 Study the section of Discussion of the lab report in 1.1. Figure out the part in this section which matches each description as followed:

A. To explain whether the data support your hypothesis.

B. To acknowledge any anomalous data or deviations from what you expected.

C. To derive conclusions, based on your findings, about the process you're studying.

D. To relate your findings to earlier work in the same area (if you can).

E. To explore the theoretical and/or practical implications of your findings.

Discussion

The experimental results obtained above were used to calculate the work done on the system by the electric heater. To calculate this, the energy balance equation, Equation 1, is used.

$$\frac{dE}{dt} = \dot{Q} - \dot{W} + \sum_{in} \dot{m}_{in}\left(h_{in} + \frac{V_{in}^2}{2} + gz_{in}\right) - \sum_{out} \dot{m}_{out}\left(h_{out} + \frac{V_{out}^2}{2} + gz_{out}\right)$$

Equation 4

When we assume that the system is at steady state, there is no heat loss, and that velocity and potential changes are negligible, the equation reduces to Equation 5.

$$\dot{W} = \dot{m}(h_{in} - h_{out})$$

Equation 5

The equation can be further simplified by making the assumption that there will be relatively small changes in pressure and temperature over the system, so enthalpy can approximated using Equation 6.

$$h_{in} - h_{out} = c_{p,avg} \cdot (T_{in} - T_{out})$$

Equation 6

In this case, an average temperature of 60 ℉ will be used to get the average specific heat of water, $c_{p,avg} = 0.999 \frac{Btu}{lbm \cdot ℉}$. By substituting mass flow for density (ρ) times volumetric flowrate ($\dot{v}$) we get our principle analytical equation, Equation 7.

$$\dot{W} = \rho \cdot \dot{V} \cdot c_{p,avg} \cdot (T_{in} - T_{out})$$

Equation 7

For this experiment, the density of water is assumed to be 62.4 lbm/ft^3 with no uncertainty. The input water temp, Tin, was found to be 46.1 ±0.4 ℉. To solve for the work put into the system, we perform a linear regression of temperature increase versus the inverse of volumetric flowrate. When this is performed the slope (m) is an average of all data points and can directly calculate the work.

$$m = \frac{\dot{W}}{\rho \cdot c_{p,avg}}$$

Equation 8

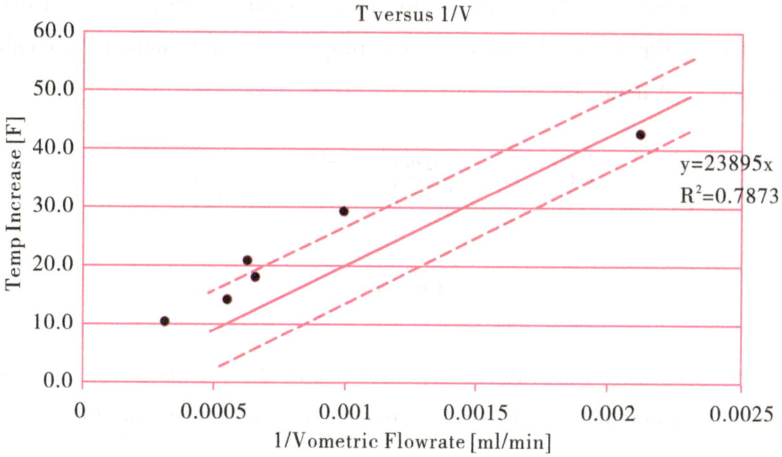

Figure 3

The confidence intervals were plotted using Equation 2, replacing the standard deviation of the sample with the standard error of the regression analysis. Uncertainty in the slope of the regression line was calculated with the same equation, but using the standard error of the slope from the regression analysis. Figure 3 shows a loosely fitting line, with $R^2 = 0.7873$ indicating the presence of other factors we have not considered.

The last data point, at 1/V of 0.002 min/mL deviated considerably form the linear model and might be considered an outlier upon further analysis.

The resulting slope and uncertainty of the regression line is $m = 23900 \pm 2500$ $\Delta T \cdot mL/min$.

While the work can be solved for by simply substituting the slope into Equation 8, the uncertainty in the work needs to be propagated.

$$\Delta \dot{W} = \dot{W} \cdot \sqrt{\left(\frac{\Delta m}{m}\right)^2 + \left(\frac{\Delta \rho}{\rho}\right)^2 + \left(\frac{\Delta c_{p,avg}}{c_{p,avg}}\right)^2}$$

Figure 9

It is assumed that the density of the fluid and specific heat are without error, so Equation 9 reduces to an equality of the percent errors of both the slope and work. To account for the bias errors from measurements, those errors are propagated in Equation 10, which is the propagation of Equation 7, combined with Equation 9 using RSS.

$$\Delta \dot{W} = \dot{W} \cdot \sqrt{\left(\frac{\Delta m}{m}\right)^2 + \left(\frac{\Delta \dot{V}}{\dot{V}}\right)^2 + \left(\frac{\Delta T}{T}\right)^2}$$

Figure 10

The uncertainty in volumetric flowrate and temperature measurements are found in the equipment table. After substituting values, we solved for the total work done on the system.

$$\dot{W} = 920 \pm 130\mathrm{W}$$

This value seems reasonable and near our predicted value.

Question:

What tenses are used in this section? Why?

Reflection

1. What is the order in which the components make up a coherent Discussion section?
2. How does this section relate to the previous sections?

1.6 Read the section of Conclusion, a summary of the results and discussion from the report in 1.3. Find out what is required in this section by matching the numbers and the components.

Conclusion

1 The first law of thermodynamics was used to calculate the work put into the system through the electric element, which was 920 ± 130W. 2 The model used in this experiment was simplistic and could account for the large uncertainty in this measurement. Figure 3 showed a large uncertainty in data when performing a linear fit, with a low R^2 value and some questionable data points. 3 A shortcoming of this model might be the assumption of an adiabatic system. The heater was constructed only of plastic and became warm during experimentation. This would indicate heat being lost to the environment, which was excluded. Inclusion of such a term would increase the work being performed by the electric heater into the system.

4 Further experimentation could be done to measure the heat loss by the system. An ammeter could be connected to the electric heating element to measure the work going in, and a similar method, like used here, could solve for the heat loss to the surroundings.

S1	A. answer to the report question
S2	B. errors discovered during experimentation
S3	C. future ideas
S4	D. concise summary of key findings

Question:

What tenses are used in this section? Why?

Reflection

What are the different uses of this section and the Discussion section in writing?

Task 2 Practicing what you have learned in Task 1.

Visit the website in Reference 47 and analyze the lab report on it from the aspects of structure and language.

Section 4

Writing in Academic Context

Unit 1

What Academic Language Is

Learning objectives

- Understanding what academic language is

What is ACADEMIC language?

"Academic language refers to the specialized vocabulary, grammar, discourse/textual, and functional skills associated with academic instruction and mastery of academic materials and tasks".

(***Source***: Saunders W., Goldenberg C., & Marcelletti, D. (2013.) English Language Development. American Educator, (Summer, 2013), 13 – 39. Retrieved from https://www.aft.org/perio dical/america n-ed ucato r/summer-2013/englis h-lan gua ge-de velo pm ent)

Task 1 Learning what specialized vocabulary is

1.1 Reread Reference 22 again.

1.2 Carefully read the vocabulary selected from the abstract in this paper and figure out the definition of the two types of vocabulary.

Academic Vocabulary	Technical Terms
· The **field** of rheology has **seen a wider application** in the food industry recently although, it is **a complex concept** and that most food systems **possess non-ideal characteristics**. · Nevertheless, the rheological behavior of foods are able to **be determined using** various **techniques and equipment**.	· The field of **rheology** has seen a wider application in **the food industry** recently although, it is a complex concept and that most **food systems** possess non-ideal characteristics. · Nevertheless, the **rheological behavior** of foods are able to be determined using various techniques and equipment.

Definitions of the two types of vocabulary:

- Academic vocabulary ______________________________

· Technical terms ____________________

1.3 List ten technical terms related to your major, both in English and in Chinese.

Task 2 Figuring out how to properly use tenses

2.1 With different colored pens mark out the tenses used in the part of abstract, introduction and conclusion in Reference 22.

2.2 Discuss the usage of tenses in writing a paper.

Task 3 Recognizing some common mistakes in academic writing

Selekted Riting Wrules

Okay—so we don't always have to be serious. It's about time I found a good use for some of the Internet jokes that appear on the e-mail in box.

1. Verbs HAS to agree with their subjects.
2. Prepositions are not words to end sentences with.
3. And don't start a sentence with a conjunction.
4. It is wrong to ever split an infinitive.
5. Avoid cliches like the plague. (They're old hat.)
6. Also, always avoid annoying alliteration.
7. Be more or less specific.
8. Parenthetical remarks (however relevant) are (usually) unnecessary.
9. Also too, never, ever use repetitive redundancies.
10. No sentence fragments.
11. Contractions aren't necessary and shouldn't be used.
12. Foreign words and phrases are not apropos.
13. Do not be redundant; do not use more words than necessary; it's highly superfluous.
14. One should NEVER generalize.
15. Comparisons are as bad as cliches.
16. Eschew ampersands & abbreviations, etc.

17. One-word sentences? Eliminate.

18. Analogies in writing are like feathers on a snake.

19. The passive voice is to be ignored.

20. Eliminate commas, that are, not necessary. Parenthetical words however should be enclosed in commas.

21. Never use a big word when a diminutive one would suffice.

22. Use words correctly, irregardless of how others use them.

23. Understatement is always the absolute best way to put forth earth-shaking ideas.

24. Eliminate quotations. As Ralph Waldo Emerson said, "I hate quotations. Tell me what you know."

25. If you've heard it once, you've heard it a thousand times: Resist hyperbole; not one writer in a million can use it correctly.

26. Puns are for children, not groan readers.

27. Go around the barn at high noon to avoid colloquialisms.

28. Even if a mixed metaphor sings, it should be derailed.

29. Who needs rhetorical questions?

30. Exaggeration is a billion times worse than understatement.

And the last one...

31. Proofread carefully to see if you have any words out.

(***Source***: http://www.ruf.rice.edu/~bioslabs/tools/report/reporterror.html.)

3.1 Read the following excerpt figure out the mistake in each sentence and categorize the mistakes in the following chart.

Categories of the mistakes

Task 4 Recognizing the academic discourse in written and oral context

4.1 Read the excerpt and try to understand the definition of "discourse or text" in linguistics.

In linguistics, the term "discourse" refers to a structural unit larger than the sentence. Of the many definitions for it in a standard dictionary, linguistics picks out length and coherence as criterial. Discourse minimally involves more than one sentence, and the sentences must be contingent. Just as every string of words is not a sentence, not every sequence of utterances is considered a "text." For discourse, there are requirements of relevance in form and especially in meaning. Texts can be created by more than one participant, as in conversation, or in various forms of monologue, most notably narrative and exposition.

4.2 Browse the website in Reference 48 to understand what "phraseological 'nuts and bolts' of writing" are and their uses in academic writing.

4.3 Reread the paper in Reference 22 and collect the "phraseological 'nuts and bolts' of writing" related to the main parts/sections of the REVIEW paper.

Sections	Phraseological "Nuts and Bolts"
Introduction	

(To be continued on next page)

(continued)

Sections		Phraseological "Nuts and Bolts"
Body	**Theme 1** Rheological Measurements of Dough and Gluten	
	Theme 2 Equipment for Rheological Measurements of Dough and Gluten	
Conclusion		

4.4 The table below is about discussion strategies. Write down the "phraseological 'nuts and bolts'" of discussion strategies as you know. Some examples are given.

Discussion Strategies	Expressions
Expressions for the Facilitator/Encourager · **Facilitator/Encourager**: This student gets discussion moving and keeps it moving, often by asking the other group members questions, sometimes about what they've just been saying.	· **Expressions for *starting a discussion*** Good morning, every one. We need to discuss/determine/find out... · **Expressions for *encouraging*** X, do you want to start? · **Expressions for *closing a discussion***
Showing your interest in the discussion · Make eye contact with the speaker. · Nod your head when you feel some points are important, or you agree with the speaker. But don't nod your head too much. · Write down the ideas you think interesting or important.	· **Expressions for *showing your interest in the discussion***
Entering or interrupting the discussion	· **Expressions for *entering or interrupt the discussion***
Giving opinions	· **Expressions *for giving opinions***

(To be continued on next page)

(continued)

Discussion Strategies	Expressions
Disagreeing politely Whenever you show your different points in discussion, please remember that you have to be polite because you respect your partners, and want your discussion continue in a friendly atmosphere.	· **Expressions for *disagreeing politely***
Checking that people are following	· **Expressions for *checking that people are following***
Checking that you have understood your partner's ideas	· **Expressions for *checking that you have understood*** So you mean that...
Making clarification	· **Expressions for *making clarification***
Changing the subject or moving on	· **Expressions for *changing the subject or* moving on**

4.5 Check your work in your group of 3 – 4 members.

Reflection

What does the term "academic discourse" mean?

Task 5 Recognizing the discourse markers

5.1 Read the excerpt and try to understand the term "discourse markers".

"Discourse markers form a group of linguistic expressions that are inseparable from discourse and fulfil important functions in spoken and written discourse interpretation. As Schiffrin (1987, 49) points out, "the analysis of discourse markers is a part of the more general analysis of discourse coherence" which is always associated with discourse cohesion. While cohesion is represented by formal linking signals in text, coherence is the underlying relations that hold between the propositions of a text on the one hand, and relations between text and context, on the other hand (Kohlani 2010, 22). Every coherent text has some sort of structure and its communicative purpose. "The communicative event which is characterized by a set of communicative purposes" is called a genre (Swales, 1990, 58). The concept of genre is more effective in representing that theoretical construct which intervenes between language function and language form."

(***Source***: http://talpykla.elaba.lt/elaba-fedora/objects/elaba:8645298/datastreams/MAIN/content)

5.2 The bold-typed expressions in the passage below are discourse markers. Figure out the use of each discourse marker.

Passage	Purposes
Once upon a time, there was a boy called Tom. He lived on a hill and picked berries for a living. He would save some berries for himself and his family and sell the rest to a fruit seller in a nearby town. The fruit seller was very happy with Tom because he would bring him a wide variety of berries. **For instance**, he would bring him strawberries, cherries, raspberries, blackberries, blueberries and mulberries. **Furthermore**, he would throw away the rotten ones and wash and clean the good ones before giving them to the fruit seller. **Therefore**, as a reward the fruit seller would give Tom one dozen bananas and mangoes each month for free. As Tom grew older he grew tired of picking berries. He wanted to become a woodcutter like his father. **However**, his father insisted that Tom continued to do his old work for some time. Tom had two elder brothers, Jack and Mark. Jack was a cobbler in the town, whereas Mark made bread in a bakery there. **Similarly**, Tom's mother also worked as a seamstress at the tailoring shop in the town.	***E. g.*** *To offer an example*

(To be continued on next page)

(continued)

Passage	Purposes
Days went by as Tom's urge to pick up an axe grew stronger. One morning Tom hid behind his house and waited for his father, mother and brothers to leave for work. When everyone left, he went inside, got his father's spare axe and walked into the woods. He came upon a small tree which he thought would be easy to cut, but just as he swung the axe, it flew from his hands and hit a bird that was perched on the lower branch of the tree. The bird was badly injured and started to bleed. Tom froze in shock at what had happened. **Meanwhile**, his father, who had forgotten his lunch at home and was walking back to get it, saw him standing like a statue in the nearby woods. He approached the scene and quickly assessed what had happened. He took out his handkerchief, wrapped the bird in it and rushed it to the stable, where the town veterinarian worked. When Tom's father returned home that evening, he was quiet but upset. He couldn't believe his son had disobeyed him like this. **Nonetheless**, he called Tom before going to bed and explained why he had not allowed him to cut wood. He told him that he was neither strong enough nor ready for such a responsibility at that age. He promised that if Tom was patient for just a few more years, he would teach him to cut wood himself. **In the end**, Tom reflected on his actions as he went to sleep that night and decided that he would rather wait for a while and be his father's woodcutting partner than be hasty and hurt his father, as well as the animals in the woods.	

5.3 Figure out the types/uses of discourse markers in the table below.

Discourse Markers	Types
also, moreover, furthermore, additionally, besides, in addition	**E. g.** For adding
similarly, likewise, in the same way	For ________
on the whole, in general, broadly speaking, as a rule, in most cases	For ________
therefore, thus, consequently, hence, as a result	For ________
however, although, whereas, despite this fact, on one hand, on the other hand, on the contrary, still, nonetheless, instead, alternatively, in contrast	For ________

(To be continued on next page)

(continued)

Discourse Markers	Types
in the past, not so long ago, recently	For ______
firstly, first of all, in the first place, to begin with, secondly, thirdly, finally	For ______
first, at first, to begin with, in the beginning, once upon a time, subsequently, earlier, meanwhile, later, afterwards, finally, at last	For ______
above all, specially, in particular, specifically, as a matter of fact, more importantly	For ______
again and again, over and over, once again, as stated	For ______
for example, for instance, such as, namely, in other words	For ______
in conclusion, finally, to sum it up, in the end, lastly, in short, eventually	For ______

5.4 Work in a group of 3 – 4 members and discuss the question: Do you think all the discourse markers in 5.3 are effective in making a presentation? Why?

5.5 Continue the discussion in your group to figure out proper discourse markers for making a presentation.

In your presentation, if you want

- to initiate discourse, including claiming the attention of the hearer (opening frame marker)
- to close discourse (closing frame marker)
- to aid the speaker in acquiring or relinquishing the floor (turn takers in group presentation)
- to serve as filler or delaying tactic used to sustain discourse or hold the floor (fillers)
- to indicate a new topic or a partial shift in topic (topic switchers)
- to denote either new or old information (information indicators)
- to mark sequential dependence (sequence/relevance markers)
- to repair one's own or other's discourse (repair markers)

Reflection

What does "discourse/textual, and functional skills" mean?

Unit 2

How to Write in Academic Context

Learning objectives

- Understanding what academic writing is
- Learning the process of academic writing
- Learning how to plan academic writing
- Learning how to revise academic writing
- Learning the features of academic writing

Task 1 Understanding what academic writing is

1.1 The whole class will be divided into 4 groups.

1.2 Students in groups 1&2 visit and read the website in Reference 49 and note down the important information you find. This is an individual job.

1.3 Students in groups 3&4 visit and read the website in Reference 50 and note down the important information you find. This is an individual job.

1.4 Each group prepares for a group presentation about what have been learned from the website.

1.5 Two groups work together and share the presentations.

Task 2 Learning the process of academic writing

2.1 Complete the description of writing steps with words provided in the box below.

editing	brainstorming	revising
publishing	drafting	organizing

To write effective essays and research papers, strong writers often use process writing.

Before you sit down to start your draft, you need to find your idea. Two of the most popular methods of fleshing out your idea are free writing and (1) ____________________. Free writing means writing every idea that comes into your head. Do not stop to edit your mistakes, just let the ideas flow. Or, try a quick list or a word map: write your idea in the center of the page and work outwards in all of the different directions you can take your story.

After you have gathered a lot of ideas, it's time to (2) __________ them. Sort through the ideas and choose which ones you want to use. Number them from the most important to the least important. Then work out an outline. More detailed your outline is, more quickly your first draft will go and more organized it will stay.

Now you have your plan and you're ready to start (3) __________ your essay or paper. Remember, this is your first rough draft. Don't worry if you stray off topic in places. Think of this stage as a free writing exercise, just with more direction. Identify the best time and location to write and eliminate potential distractions. Make writing a regular part of your day.

Your paper can change a great deal during this stage. When (4) ____________________ your work, you will either add more relevant information or eliminate any sentences that don't quite fit, and rearrange the flow and sequence of information. The most effective way to revise your work is to ask for a second opinion. Ask friends or classmates to take a look and give you feedback, and if something isn't working rewrite it and replace it.

Proofreading, also known as (5) __________, refers to fining tune your paper line by line. Check for repetition, clarity, grammar, spelling and punctuation. Editing is an extremely detailed process and its best when performed by a professional.

Finally, you have a completed manuscript ready for (6) __________. Professional writers will have their work printed in a newspaper or a magazine or posted online. For students, it usually means submitting the work for a grade.

(***Source***: Writing Research Papers: From Essay to Research paper, Macmillan Publishers Limited by Dorothy E. Zemach, Daniel Broudy, Chris Valvona, 2011)

2.2 Work in pairs to present an oral summary on writing process. Think about what discourse markers can help express yourself clearly.

Writing Process

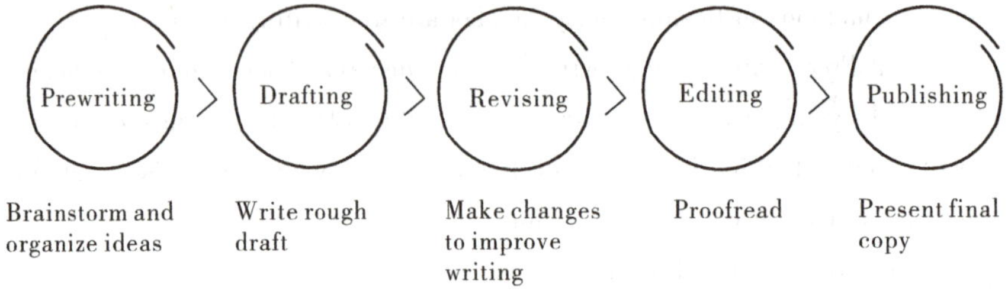

Task 3 Learning how to plan academic writing

3.1 Read the tips of planning academic writing.

Tips:

- Make a realistic time table, working back from your deadline.
- Plan your reading and research. How long will it take to gather and organize the information you need?
- Develop a writing plan, perhaps using headings or mind maps.

3.2 Develop your writing plan for your review paper on academic integrity. Pay attention to the deadline assigned by your teacher.

Task 4 Learning how to revise academic writing

4.1 Read the tips of revising academic writing.

Tips:

- First of all, try to put your writing away and leave it for a day or so. This means you can pick it up and look at it with a fresh eye. It also allows your brain to mull over the ideas subconsciously, which can really help.
- When you return to your writing to review it, you need to proofread for content, and also for the writing, for example spelling, grammar, punctuation.
- You can't do both at the same time. If you are concentrating on the content and whether everything makes sense, you won't notice spelling mistakes. So

you need to read the writing for the content and the writing separately.

- Check overall organization: tone of expression, logical flow of introduction, coherence and depth of discussion in body, effectiveness of conclusion.
- Paragraph level concerns: topic sentences, sequence of ideas within paragraphs, use of details to support generalizations, summary sentences where necessary, use of transitions within and between paragraphs.
- Sentence level concerns: sentence structure, word choices, punctuation, spelling.
- Documentation: consistent use of one system, citation of all material not considered common knowledge, appropriate use of endnotes or footnotes, accuracy of list of works cited.

· It's a good idea to ask someone to proofread your writing. They can often pick up errors you have missed, because you're too close to it.

4.2 Revise your review paper.

Task 5 Learning the features of academic writing

5.1 Browse the websites in Reference 51 and Reference 52, and note down the important information.

5.2 Read the two abstracts below and answer the questions.

· **Abstract 1:**

My purpose in this paper is to complicate the genealogies of the concept of culture as a way of life that have held sway within cultural studies. I do so by reviewing key aspects in the development of this concept within the "Americanist" tradition of anthropology pioneered by Franz Boas in the opening decades of the twentieth century and continued by a generation of Boas's students including Ruth Benedict, Alfred Kroeber and Margaret Mead. I focus on three issues: the respects in which the "culture concept" was shaped by aesthetic conceptions of form, its spatial registers, and its functioning as a new surface of government, partially displacing that of race, in the development of American multicultural policies in the 1920s and 1930s. In relating these concerns to Graeme Turner's enduring interest in the processes through which culture is "made national", I indicate how the spatial registers of the culture concept anticipate contemporary approaches to these questions. I also outline what Australian cultural studies has to learn from the American evolution of the culture concept in view of the respects in which the latter was shaped by the racial dynamics of a "settler" society during a period of heightened immigration from new sources. In concluding, I review the broader implications of the fusion of aesthetic and anthropological forms of expertise that informed the development of the culture concept.

(***Source***: Tony Bennett (2015): Cultural Studies and the Culture Concept, Cultural Studies, DOI: 10.1080/09502386.2014.1000605, retrieved from http://dx.doi.org/10.1080/09502386.2014.1000605)

· **Abstract 2:**

Background

In developed countries, smoking is associated with increased risk of diabetes. Little is known about the association in China, where cigarette consumption has increased (first in urban, then in rural areas) relatively recently. Moreover, uncertainty remains about the effect of smoking cessation on diabetes in China and elsewhere. We aimed to assess the associations of smoking and smoking cessation with risk of incident diabetes among Chinese adults.

Methods

The prospective China Kadoorie Biobank enrolled 512, 891 adults (59% women) aged 30 – 79 years during 2004 – 2008 from ten diverse areas (five urban and five rural) across China. Participants were interviewed at study assessment clinics, underwent physical measurements, and had a non-fasting blood sample taken. Participants were separated into four categories according to smoking history: never-smokers, ever-regular smokers, ex-smokers, and occasional smokers. Incident diabetes cases were identified through linkage with diabetes surveillance systems, the national health insurance system, and death registries. All analyses were done separately in men and women and Cox regression was used to yield adjusted hazards ratios (HRs) for diabetes associated with smoking.

Findings

68% (n = 134975) of men ever smoked regularly compared with 3% (n = 7811) of women. During 9 years' follow-up, 13,652 new-onset diabetes cases were recorded among 482589 participants without previous diabetes. Among urban men, smokers had an adjusted HR of 1·18 (95% CI 1·12–1·25) for diabetes. HRs increased with younger age at first smoking regularly (1·12, 1·20, and 1·27 at ≥25 years, 20–24 years, and <20 years, respectively; p for trend = 0·00073) and with greater amount smoked (1·11, 1·15, 1·42, and 1·63 for <20, 20–29, 30–39 and ≥40 cigarettes per day; p for trend <0·0001). Among rural men, similar, albeit more modest, associations were seen. Overall, HRs were more extreme at higher levels of adiposity. Among men who stopped by choice, there was no excess risk within 5 years of cessation, contrasting with those who stopped because of illness (0·92 [0·75–1·12] vs 1·42 [1·23–1·63]). Among the few women who ever smoked regularly, the excess risk of diabetes was significant (1·33 [1·20–1·47]).

Interpretation

Among Chinese adults, smoking was associated with increased risk of diabetes, with no significant excess risk following voluntary smoking cessation.

(***Source***: https://www.sciencedirect.com/science/article/pii/S2468266718300264? via%3Dihub)

Questions:

1. Why are different persons used in these two abstracts?
2. Why is passive voice used in Abstract 2?
3. Which one sounds objective? Why?
4. How do you understand the objectivity in academic writing?
5. How do you keep objective tone in your academic writing?

5.3 Discuss the questions in 5.2 in your group of 3–4 members.

5.4 Read the tips below and understand how to express oneself certainly or cautiously.

➲ Tips:

Academics use different words to express how certain or cautious they are about claims/point of view.

· When they are more certain, they **boost** their statements to be assertive about their claims.

E. g. *Research* ***claims*** *that high consumption of fizzy drinks containing sugar* ***contributes*** *to the development of type II diabetes.*

· When being careful, they **hedge** their statements.

E. g. *Research* ***suggests*** *that high consumption of fizzy drinks containing sugar* ***may*** *contribute to the development of type II diabetes.*

5.5 Work in your group and change the boosting/assertive statement into different hedging ways listed in the table below.

Ways of Hedging	Expressions for Hedging	Assertive Statement: *Driving at night is dangerous.*
Lexical verbs (i. e. words which carry the attitude or intention of the writer)	suggest, hint, indicate, believe, assume, recommend, tend to	
Full modal verbs (i. e. verbs of attitude and meaning + the bare infinitive e. g. can claim)	may, might, could, can, etc.	
Compounds (including semi-modal verbs followed by the infinitive "to", e. g. ought to and multiword compounds, e. g. be likely to)	ought to, have to, dare to seem, appear be able to, be unable to, be likely to, be unlikely to	
Adverbs (including frequency adverbs/adverbial phrases)	possibly, probably, approximately, presumably sometimes, seldom, always, often	
Adjectives/prepositional phrases	reasonable, possible, probable, likely in all likelihood, to some extent/to a certain extent, with caution	
Quantifiers/quantifying phrases	some, a few, many, somewhat, the majority of	

5.6 Watch Video 4 about the use of abstract nouns in academic writing and take notes.

5.7 Discuss the question below in your group of 3 – 4 members. You can refer to the notes taken in 5.6.

Question: On what occasions in academic writing can we use the abstract nouns?

5.8 Read the three pieces of writing under the same topic and decide which one is academic writing and explain why.

Writing 1:

I think that essay writing is an important skill for all of us students. Don't you see how many marks are given for this? Lots of students agree that they are marooned if they can't write a decent essay. In my opinion (as a struggling student), we should have lessons in essay writing from day one!!!

Writing 2:

It is in fact correct to say that academic essay writing is of utmost importance in the attainment of a university degree. A high proportion of marks are allocated to the compilation of essay assignments as part of a university course to the point where it could be the causation of terminating a degree program because of failure. There is somewhat of an obligation for universities in the provision of services to the student population to educate their students in the intricacies of essay writing early in their undergraduate first year.

Writing 3:

Essay writing is an important skill for tertiary students. Academic essays can attract a considerable proportion of assessment marks in most degree programs. Therefore, students may require a firm grounding in academic essay writing skills at the start of their first year to assist them to succeed in their university studies.

5.9 Work in your group and discuss your answers to 5.8.

5.10 Work in your group and fill in the blanks below.

➲ **How to write in an objective tone**

Never use contractions

__________ use colloquialisms

Avoid __________ use of "I", "you" & "we"
Avoid __________ use of the passive voice
Avoid __________ use of hedging
Avoid __________ use of nominalization
Avoid __________ use of complicated words and sentence structure

5.11 Read the excerpt from the book *Elements of Style* written by William Strunk Jr. to understand how to keep concise in academic writing.

> Vigorous writing is concise. A sentence should contain no unnecessary words, a paragraph no unnecessary sentences, for the same reason that a drawing should have no unnecessary lines and a machine no unnecessary parts.

5.12 Make the sentences below concise.

1. Unencumbered by a sense of responsibility, Jason left his wife with forty-nine kids and a can of beans.

   ____

   ____

2. Smith College, which was founded in 1871, is the premier all-women's college in the United States.

   ____

   ____

3. There are twenty-five students who have already expressed a desire to attend the program next summer.

   ____

   ____

5.13 Check your work in your group and explain your improvements.

5.14 Read the two sentences below and figure out which one is precise.

1. Chemists had attempted to synthesize quinine for the previous hundred years but all they had achieved was to discover the extreme complexity of the problem.
2. The volatile oily liquid beta-chloro-beta-ethyl sulphide was first synthesized

in 1854, and in 1887 it was reported to produce blisters if it touched the skin. It was called mustard gas and was used at Ypres in 1917, when it caused many thousands of casualties.

5.15 Check your answers to 5.14 in your group.

5.16 Read the two groups of statements and figure out which one in each group is of academic style.

Group 1

1. On the surface, why women and domestic architecture were associated were obvious.
2. On the surface, the reasons for the association between women and domestic architecture were obvious.

Group 2

1. How big this group is varies in different centers.
2. The size of this group varies in different centers.

5.17 Read the two clauses below, compare them with the clause in Sentence 1, Group 1, 5.16 and think about the questions.

1. This conforms conveniently with Maslow's (1970) claim that human motivation is related to a hierarchy of human needs.

2. The other way in which the economic aspects of military expenditure were presented was in the form of the public expenditure costs.

Questions:

1. Do you think the uses of subordinate clauses in this part are proper? Why?
2. When should we use the subordinate clauses?

5.18 Work in your group and discuss your answers to 5.13.

Reflection

What do you think are the features of academic writing according to what you have learned in Task 5.

Task 6 Putting it together

6.1 The following passage is the beginning part of a paper. Rewrite it in a more academic style.

ICT technologies are expected to hold the ignition key to the reduction of the greenhouse gases produced worldwide, which is a nondebatable global priority. The importance of "greening of the Internet", therefore, is recognized as a primary design goal of the future global network infrastructures. Indeed, the Internet today already accounts for about 2% world energy consumption, but with the current trend of shifting offline services online, this percentage will grow significantly in the next few years, and it will be pushed further by the forthcoming Internet-based platforms that require always-on connectivity. In this paper we present...

6.2 Rewrite the following passage in a more academic style.

Until only a few hundred years ago doctors didn't operate on people, it was barbers. No one had taught them and they'd never got any qualifications. They just did what they'd learned when they were learning to be barbers. Doctors had promised not to hurt anyone so they wouldn't cut people and were not even supposed to watch. But the doctors did watch if they were following the rules properly and he sat on a big chair, high up, and read out what the barber was supposed to do. He read this in Latin, which, of course, the barber didn't understand. Of course, if you died, it was always the barber's fault and if you got better, the doctor got the praise. In any case, the doctor got the most money.

6.3 Share your work in your group and explain why the improvements are made.

Section 5

Communication in Academic Context

Unit 1

How to Make Presentations

Learning objectives

- Understanding what academic presentation is
- Knowing the procedure of making an academic presentation
- Knowing how to prepare to present a presentation in academic context
- Learning how to get started
- Expressing ideas in main body of presentations with proper signposts
- Ending a presentation
- Making a good PPT presentation
- Making an effective poster presentation

Task 1 Understanding what academic presentation is

1.1 Read the excerpt from the website https://edubirdie.com/blog/presentation-topics and highlight the important information you think.

Academic presentation—what is it and how to create it?

Simply, this is a presentation done for academic purposes. It usually takes place at school or college. Your teacher or professor will pick a certain topic or ask you to choose an interesting one for you. You have to study this topic, research, and analyze it to prepare an amazing PowerPoint presentation.

An academic presentation is an effective task that develops a student's communication skills. They later use these skills in their professional and personal lives in the future after studies. Shy students might struggle with talking in front of an audience but learning about the true power of speech will help them overcome their fears. The classroom is a safe environment where students get to share their ideas and opinions with others. It later helps them to join reputable US universities and firms that only recruit the best.

1.2 Share your highlighted information in your group of 3–4 members.

Task 2 Knowing the procedure of making academic presentations

Academic presentations can be individual or in groups.

Academic presentations and seminars often take the following form:

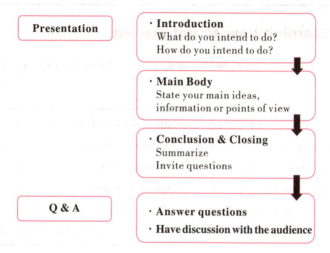

Presentation

- **Introduction**
  What do you intend to do?
  How do you intend to do?
- **Main Body**
  State your main ideas, information or points of view
- **Conclusion & Closing**
  Summarize
  Invite questions

Q & A

- **Answer questions**
- **Have discussion with the audience**

Task 3 Knowing how to prepare for presenting a presentation in academic context

It can be very boring to listen to something read aloud. Therefore, what you must do is to follow the points in the chart below.

1. **Decide on a time limit for your talk.**

   Tell your audience what it is. Stick to your time limit.

2. **Write out your spoken presentation in the way you intend to say it.**

   This means that you must do some of the work of writing the paper again, in a sense. Pay attention to the language you are going to use, for written language is different from spoken language. Do the writing job as suggested below:

   - *Concentrate only on the main points*
   - *Try to make your presentation lively and interesting*
   - *Write out everything you have to say*
   - *Reduce it to outline notes*

3. **Speak from the outline notes.**

4. Look at your audience when you are speaking.

5. Make a strong ending.

One way of doing this is to repeat your main points briefly and invite questions or points of view.

Task 4 Learning how to get started

4.1 Read the instructions below and discuss the questions in your group of 3 – 4 members.

The mission of opening your presentation is to grab the audience's interest, establish rapport and introduce your topic. Openings count a lot for a successful presentation, not only because they set the tone of your presentation, but also can maximize the immediate impact on your audience. Typically, in an organized opening, the subject, purpose and big picture or outline of the main ideas to be covered in the presentation are stated.

Questions:

1. How do you get your audience's attention?
    - Greeting
    - Telling a story
    - Quote others' words
    - Asking a question

    Other ways:

2. Do you think it is an easy job to tell a story or a joke as attention-getter? Are you a good story teller? What should you do to be a good story teller?
3. Whose words can be quoted?
4. What kind of question are good for grab audience's interest?
5. How long is a good attention getter?
6. Do we need a dramatic attention getter for our academic presentation?
7. What is the basic difference between academic presentation and public speech?

4.2 Work in your group of 3 – 4 members and discuss how to make the introduction part of the presentation on each of the following topics. You may refer to Reference 53.

- Human Evolution
- Animals and Environment
- Digital Books
- Cloud Computing and the Internet
- Genetically Modified Foods

4.3 Orally practice getting started by using the ways you have discussed in 4.2.

Task 5 Expressing ideas in main body of presentations with proper signposts

5.1 Please read the following passage and study the notes below.

To the ordinary man, one kind of oil may be as important as another. But when the politician or the engineer refers to oil he almost always means mineral oil, the oil that drives tanks, aeroplanes and warships, motor-cars and diesel locomotives; the oil that is used to lubricate all kinds of machinery. This is the oil that has changed the life of the common man. When it is refined into petrol it is used to drive the internal combustion engine. To it we owe the existence of the motorcar, which has replaced the private carriage drawn by the horse. To it we owe the possibility of flying. It has changed the methods of warfare on land and sea. This kind of oil comes out of the earth. Because it burns well, it is used as fuel and in some ways it is superior to coal in this respect. Many big ships now burn oil instead of coal. Because it burns brightly, it is used for illumination; countless homes are still illuminated with oil-burning lamps. Because it is very slippery, it is used for lubrication. Two metal surfaces rubbing together cause friction and heat; but if they are separated by a thin film of oil, the friction and heat are reduced. No machine would work for long if it were not properly lubricated. The oil used for this purpose must be of the correct thickness; if it is too thin it will not give sufficient lubrication, and if it is too thick it will not reach all parts that must be lubricated.

Notes:

Mineral oil
A. most common mineral oil
B. from earth
C. use 1. for tanks, aeroplanes and warships, motor-cars and diesel locomotives
2. to lubricate all kinds of machinery
3. owe the existence of the motorcar, possibility of flying
D. Properties 1. Burns well→fuel
2. Burns brightly→illumination
3. Slippery→lubrication

Questions:

1. What is the topic of this presentation?
2. From what aspects is the topic developed?
3. How do you logically express these aspects in your presentation?
4. How many ways do you know to organize the main ideas in the body to develop the topic and the thesis statement of your presentation? What are the ways?
5. Why do you select some certain ways to develop the topic and the thesis statement of your presentation?
6. If your presentation will last more than 20 minutes and you have many ideas to cover, do you think the expressions like "Firstly, ... Secondly, ... What's more, ... Finally, ..." are always good enough for oral communication? Why or why not?

5.2 Read Signposting in Presentations 1 in Reference 53 and Signposting in Presentations 2 in Reference 54 to learn how to well communicate your ideas with the audience.

5.3 Give a short presentation with the notes above. Pay attention to the signposting expressions you will use. Remember that you are having a communication with your audience while delivering the presentation.

5.4 Figure out the ways of organization for the topics in 4.2.

Task 6 Ending a presentation

6.1 Read the signposts below and analyze their functions in ending a presentation by matching column A with column B.

A Signposts	B Functions
1. So, to sum up... In conclusion (we can say that)... 2. If you have any questions, I'll be pleased to answer them 3. Let me end by saying... 4. I'd like to finish by emphasizing... 5. Thank you for your attention/time/listening.	A. Conclusion B. Inviting questions C. Closing

6.2 Discuss the questions in your group of 3 – 4 members

1. According to your answers in 6.1, what should you do in the end of your presentation?
2. How do you draw a meaningful and powerful conclusion instead of a hasty one? Remember conclusion is the last chance you will leave positive impression on your audience about your topic or idea.
3. Do you think the expression "That's all." is proper to end your presentation? Why?
4. What are you going to say if you fail to get the question from the audience?
5. What are you going to say if you cannot answer the question beyond your knowledge?
6. What are you going to say if your audience's question implies that he/she misunderstands your ideas?
7. What are you going to say if you realize that you don't have enough time to answer a question raised by the audience?
8. What are you going to say if you realize that the person who asks question is not friendly?
9. What are you going to say to end Q&A session?
10. What are you going to say if there is no question after your invitation?
11. What are you going to say if you have no time to end your presentation?

Task 7 Making a good PPT presentation

7.1 PowerPoint (PPT) is a commonly used presentation program, created by Robert Gaskins and Dennis Austin at a software company named Forethought, Inc. It was released on April 20, 1987. Think about the questions: Why is this program named PowerPoint? Why do we use this program when we deliver a presentation? Is it proper to use it as a script of your presentation?

7.2 Learn how to make a good slide by watching video #5. Please note down the pitfalls of a bad PPT.

7.3 Read the article "Creating a PowerPoint for an Academic Presentation" in Reference 55. You can improve this document according to the notes in 7.2 or your knowledge of making PPT.

7.4 Share your improvements in your group of 3 – 4 members.

7.5 Browse the websites in Reference 56 and 57, and watch any presentation by Steve Jobs on the Internet. Note down the important tips of making a PPT presentation.

7.6 Discuss your notes in a small group and answer the question: How can we make a good PPT-Presentation?

Task 8 Making a good poster presentation

8.1 Browse the websites in Reference 58, 59 and 60, and note down the answers to the questions below.

Questions:

1. What is poster presentation?
2. How do you make an effective poster presentation?

8.2 Be ready to make a poster presentation on the topic "What is Culture" in the next class. You are required to do some research on the term "culture" for preparing this presentation.

The detailed instructions on this activity are located in Task 1, Unit 1, Section 6.

Unit 2

How to Have a Discussion

Learning objectives

- Knowing what discussion is
- Learning to have different types of discussions

Task 1 Knowing what discussion is

1.1 Recognize the roles in a discussion

There are different **roles** you can play in a discussion:

- **Facilitator** moderates group's discussion by keeping them focused on the task/topic and ensuring that all members participate. Facilitator has to start and end the discussion.
- **Recorder/Reporter/Summarizer** keeps notes of discussion and creates a written summary of the group's discussion to share in a forum open to all. Alternatively, for non-online courses, reports orally to the class. Responsible for any written assignment submissions.
- **Cheerleader/Energizer** encourages participation of other group members through positive reactions to posting. Encourages non-posters to participate through direct interaction with them.
- **Initiator** is responsible for the first post in each discussion assignment.
- **Posters and Lurkers** in larger group, divide students into two groups for each assignment. One week they participate; one week they "watch and listen" (aka lurking).
- **Responder** does not initiate responses to the discussion question but responds to others' posts, questioning and pushing them further in their thinking.
- **Timekeeper** monitors time/deadlines and moves group along so that they complete the task in the available time. They will remind of the time during the discussion.
- **Checker** makes sure that all group members understand the concepts and the group's conclusions.

- **Elaborator** relates the discussion with prior concepts and knowledge.
- **Research-Runner** finds additional material as needed and is the liaison between the group and the instructor.
- **Skeptic** helps the group avoid coming to agreement too quickly; makes sure all possibilities and alternatives are explored.

1.2 Knowing types of discussion

There are a variety of different types of discussions that occur naturally and which we can recreate in the classroom. These include discussions where the participants have to:

- Make decisions (e. g. decide who to invite to a party and where to seat them).
- Give and/or share their opinions on a given topic (e. g. discussing beliefs about the effectiveness of capital punishment).
- Create something (e. g. plan and make a poster as a medium for feedback on a language course).
- Solve a problem (e. g. discussing the situations behind a series of logic problems).

1.3 Read the websites in Reference 61, 62 and 63, and notes down important information related to discussion skills.

Task 2 Learning to have discussions

2.1 Read the passage below and answer the questions individually.

Brainstorming

Brainstorming is a group creativity technique by which efforts are made to find a conclusion for a specific problem by gathering a list of ideas spontaneously contributed by its members. In other words, brainstorming is a situation where a group of people meet to generate new ideas and solutions around a specific domain of interest by removing inhibitions. People are able to think more freely and they suggest many spontaneous new ideas as possible. All the ideas are noted down and are not criticized and after brainstorming session the ideas are evaluated. The term was popularized by Alex Faickney Osborn in the 1953 book *Applied Imagination*.

Origin

Advertising executive Alex F. Osborn began developing methods for creative problem-solving in 1939. He was frustrated by employees' inability to develop creative ideas individually for ad campaigns. In response, he began hosting group-thinking sessions and discovered a significant improvement in the quality and quantity of ideas produced by employees. During the period when Osborn made his concept, he started writing on creative thinking, and the first notable book where he mentioned the term brainstorming is "How to Think Up" in 1942. Osborn outlined his method in the 1948 book *Your Creative Power* in chapter 33, "How to Organize a Squad to Create Ideas".

Osborn's Method

Osborn claimed that two principles contribute to "ideative efficacy", these being "Defer judgment" and "Reach for quantity". Following these two principles were his four general rules of brainstorming, established with intention to reduce social inhibitions among group members, stimulate idea generation, and increase overall creativity of the group.

· Focus on quantity: This rule is a means of enhancing divergent production, aiming to facilitate problem solving through the maxim quantity breeds quality. The assumption is that the greater the number of ideas generate the bigger the chance of producing a radical and effective solution.

· Withhold criticism: In brainstorming, criticism of ideas generated should be put "on hold". Instead, participants should focus on extending or adding to ideas, reserving criticism for a later "critical stage" of the process. By suspending judgment, participants will feel free to generate unusual ideas.

· Welcome unusual ideas: To get a good long list of suggestions, wild ideas are encouraged. They can be generated by looking from new perspectives and suspending assumptions. These new ways of thinking might give you better solutions.

· Combine and improve ideas: As suggested by the slogan "1 + 1 = 3". It is believed to stimulate the building of ideas by a process of association.

Applications

Osborn notes that brainstorming should address a specific question; he held that sessions addressing multiple questions were inefficient. Further, the problem must require the generation of ideas rather than judgment; he uses examples such as generating possible names for a product as proper brainstorming material, whereas analytical judgments such as whether or not to marry do not have any need for brainstorming.

Brainstorming Groups

Osborn envisioned groups of around 12 participants, including both experts and novices. Participants are encouraged to provide wild and unexpected answers. Ideas receive no criticism or discussion. The group simply provides ideas that might lead to a solution and apply no analytical judgment as to the feasibility. The judgments are reserved for a later date.

Incentives and Brainstorming

Some research indicates that incentives can augment creative processes. Participants were divided into three conditions. In Condition Ⅰ, a flat fee was paid to all participants. In the Condition Ⅱ, participants were awarded points for every unique idea of their own, and subjects were paid for the points that they earned. In Condition Ⅲ, subjects were paid based on the impact that their idea had on the group; this was measured by counting the number of group ideas derived from the specific subject's ideas. Condition Ⅲ outperformed Condition Ⅱ, and Condition Ⅱ Outperformed Condition Ⅰ at a statistically significant level for most measures. The results demonstrated that participants were willing to work far longer to achieve unique results in the expectation of compensation.

Some research claims to refute Osborn's claim that group brainstorming could generate more ideas than individuals working alone. Research from Michael Diehl and Wolfgang Stroebe demonstrated that groups brainstorming together produce fewer ideas than individuals working separately. Their conclusions were based on a review of 22 other studies, 18 of which corroborated their findings.

(***Source***: https://en.wikipedia.org/wiki/Brainstorming)

Questions:

1. What is brainstorming?
2. What is the origin of brainstorming?
3. According to Osborn, what are the two principles contributing to ideative efficacy?
4. What are the four general rules of brainstorming?

5. What kind of problems call for brainstorming?
6. What is the function of the brainstorming group?
7. What are the three incentive conditions of brainstorming?
8. Why is the third incentive condition outperform the other two?

2.2 Work in your group, and discuss your answers to the questions in 2.1.

- You have to work in groups of 4 members.
- Each member will take the different roles: facilitator, recorder, checker or timekeeper.
- Before discussion, your group has to decide the type of presentation.
- Do brainstorming when collecting the answers to each question from your group members.
- When the discussion is finished, every group should hand in a memo of the discussion results. You have to use *Template of Memo* in Appendix Ⅱ.
- The time for discussion is 30 minutes.
- While you are having the discussion, your teacher will observe it. If you fail to play your role, you will be punished.
- You may refer to *Strategic Questioning Techniques for Discussion* in Reference 64.

2.3 Read the passage below to understand the thinking skill "putting on thinking hats".

Putting on Your Thinking Hats

Brainstorming is one of the most important and widely used means to innovate. However, brainstorming sessions are often not as productive as expected due to ego clashes, arguments and lack of focus. A leader in creative thinking, Edward De Bono, developed the "six thinking hats technique" that overcomes most of the pitfalls of regular brainstorming. The technique proposes and explains six directions of thought. Wearing only one hat at any given point in time helps an individual (or a team) focus on various aspects of the topic of discussion in a non-confrontational manner, thereby creating considerable synergy and enabling the brainstorming to be much more fruitful and productive. Furthermore, the use of lateral thinking triggers and catalyzes creative thinking and expands the scope of both the problems, as well as the solutions generated.

Why Six Thinking Hats?

Human brains think along multiple directions at the same time. This way of thinking is extremely efficient when quick decisions need to be made. The brain reduces the complexity of data through fuzzy logic or "bucketization". However, when a problem or issue needs to be thought through in detail, the brain needs to resist fixing on a pattern too quickly. This is notoriously difficult to achieve, even at an individual level. In a team brainstorming session, this problem worsens. Solutions can be quickly jumped even while the problem is not fully understood, ideas can be shot down even before being fully developed and solutions can be pointed out as infeasible without the opportunity to consider them. De Bono proposed the separation of directions of thought into the six thinking hats. By forcing one or more brains to think along one direction at one time, most collisions are avoided and synergy is considerably increased. The technique also encourages the use of lateral thinking approaches leading to increased brainstorming productivity.

White Hat—Data, Facts and Figures

While wearing the white hat, all available data is put forth. Data can be both within (and outside) the scope of the discussion, effectively deferring judgment of data. There is a focus on neutral facts and figures. Participants can include opinions of others, where the opinion becomes the fact irrespective of whether the opinion is believed to be accurate or not. It is important to be specific about data to reduce ambiguity as much as possible. Overall, the objective is to facilitate a deeper understanding of the breadth and depth of the relevant issue with the neutral deposition of data on the table.

Red Hat—Emotions

The red hat is the hat of emotions. Emotions can be positive (happiness, joy, wonder, enthusiasm, hope, expectation), negative (anger, disappointment mental-blocks, jealousy, cynicism) or neutral (intuition, complex emotional judgment, curiosity). It is important to be free from obligations to effectively wear the red hat. The objectives of the red hat are to lend credibility to "emotional" thinking, help feeling surface (rather than remain hidden yet play a major part in the discussion) and free the participants from having to justify emotions, complex judgment or intuitive thoughts. This adds a new dimension to both problems and solutions.

Green Hat—Ideas

The green hat is the hat of ideas. By removing judgment and feasibility analysis from the scope of the discussion, participants are free to generate crazy ideas that may not be immediately relevant or feasible. New ideas can be constructed from other ideas generated and pooled at the table. The setting is ripe for lateral "out-of-the-box" thinking. In the green hat phase, a team consisting of individuals from varying backgrounds, age groups, cultures and having different perspectives to the issue becomes a fantastic asset, rather than a liability, to brainstorming. The objective of the green hat is to generate as many (and as varied) ideas as possible.

Yellow Hat—Positivism

This is the "be positive" hat. Positive aspects of the issue are stressed. A positive outlook toward problem solving is maintained. The focus is on "how-to-make-it-happen" rather than "how-this-may-not-work". Ideas from the common pool (the ideas are now not tied to the individuals who generated them) are picked up for further development. The positive factors are noted. All possible positive effects of an idea are noted and discussed. The objective is to be positive about "making-things-happen" and understand all the benefits of various ideas.

Black Hat—Critical Judgment

The black hat is the hat of caution and judgment. All constraints are noted down. The scope of the issue may be fine-tuned. Feasibility of ideas is evaluated. Pitfalls and shortcomings are identified. Problems and ideas are prioritized based on relevance or impact. A cautious approach is used. Participants can play the "devil's advocate", discuss worst-case scenarios and identify bottlenecks and weak links. The objective of this hat is to critically evaluate and judge issues and ideas, and converge on a specific problem or solution areas.

Blue Hat—Control and Overview

The blue hat controls the flow of the brainstorming session—what hat to wear, when and for how long. The agenda and timelines are decided. There can be a discussion on the sequence of hats to be used. Typically, the controller of the session wears the blue hat to guide the session and ensure that all participants continue to wear one hat at t time. The blue hat is used to arbitrate and re-focus. The blue hat is also used to summarize and conclude.

(***Source***: https://triz-journal.com/beyond-triz-the-world-of-systematic-innovation/)

2.4 Work in pairs and answer the questions according to the passage above.

1. Why are brainstorming sessions not as effective as expected?
2. Why is it advocated to wear only one hat at any given point in time?
3. What is the working mechanism of the "six thinking hats principle"?
4. What is the white hat concerned with?
5. Why is it important to wear a red hat?
6. What is the objective of a green hat?
7. What is stressed while wearing a yellow hat?
8. What is the objective of a black hat?
9. What is the purpose of wearing a blue hat?

2.5 Work in a group of 6 members and each member puts ONE thinking hat. Figure out a dilemma in your life, discuss this situation with your hat and make a decision.

- When the discussion is finished, every group should hand in a memo of the discussion results. The template of memo is in Reference #65.
- The time for discussion is 30 minutes.
- You may refer to the expressions for discussion in Reading Booklet.

2.6 Make a group presentation to show your discussion results.

Task 3 Learning to make a panel discussion

3.1 Read the excerpt below and highlight the important information related to panel discussion.

Panel discussions at conferences are a useful way to trigger an exchange of viewpoints among experts, either with prepared statements or in response to questions from the audience. Because they involve on-the-spot interaction, they are more difficult to prepare for than presentations. Because they may involve divergence of viewpoints and possibly competition for speaking time, they are also more difficult to manage than the normal questions at the end of a presentation. For the same reasons, they are more challenging to moderate than a regular conference session.

(*Source*: https://www.nature.com/scitable/topicpage/panel-discussions-13909630)

3.2 Watch Video 6 and notes down the divergent viewpoints given by the speakers and the questions given by the host of the panel discussion.

Your notes:

- Topic of discussion: ______________________
- Divergent viewpoints on the topic:

- Questions given by the host:

3.3 Watch Video 6 again and answer the questions below.

Q1. What is the role of a moderator in a panel?

Q2. What is the role of panelists in a panel?

3.4 Study the information on the website in Reference 65 and summarize do's and don'ts of being a moderator and panelists.

➲ Your notes:

- Do's and Don'ts of being a moderator

Do's	Don'ts

- Do's and Don'ts of being panelists

Do's	Don'ts

3.5 Study the website in Reference 65 and watch Video 7 and summarize other important information about having a panel discussion.

➲ Your notes:

- Difference between Video 6 and Video 7:

- Types of panel discussion:

- Tips of having a panel discussion:

3.6 Work in a group and do a Model Panel Discussion.

- Teacher will present a topic for the panel discussion: Can Alternative Energy Effectively Replace Fossil Fuels?
- Suppose there are 20 students in a class, three students will play a role of panel moderator. They will work together and prepare some questions for the panel discussion before the next class.
- Five students will be the audience of the panel. They will work together and prepare at least 6 questions for the panelists. The number of students as the audience can be justified according to the real numbers in different classes.
- The rest of class will be the panelists. They will be divided into 3 groups. There are 4 members in each group. First, each member will be labeled with the numbers from 1 to 4. Then, all the No. 1s & 2s will be the positive side of the topic and all the No. 3s & 4s, the negative side of the topic.
- Before the next class, each side will work out four different contentions and their relevant evidence.

- Each member will be responsible for one contention in the panel and be ready to do a 2-minute presentation before answering the questions from the panel moderator.
- Every member should be ready to answer the questions from the panel moderator.

Section 6

Developing Intercultural Communicative Competence

Unit 1

Developing Intercultural Communicative Competence

Learning objectives

- Understanding what culture is
- Understanding what culture shock is and ways to deal with it
- Understanding what intercultural communication is
- Learning how to avoid barriers in cross-cultural communication

Task 1 Learning what culture is

1.1 Do your research about the question what culture is on the website or in the library. Take notes and make a matrix.

1.2 Make a poster according to your notes and matrix.

1.3 Prepare a poster presentation.

- You have to print out the poster and bring it to class.
- Do rehearsal at least 2 times.

1.4 In class, stick your poster on the walls or boards before the poster session.

1.5 In poster session, do your own poster presentation at least one time, listen to at least 2 presentations and discuss on at least 3 posters. In this session, please take notes.

1.6 Finish a reflection journal over this poster session with the title: 3 things I've learned from the poster session.

Task 2 Learning what culture shock is and how to do with it

2.1 Do your research about the questions below. Take notes and make a matrix.

Q1: What is culture shock?

Q2: How can we do with it?

2.2 Work in a small group and share your matrix.

2.3 Write a summary of your discussion.

Task 3 Understanding what intercultural communication is

3.1 Browse the websites in Reference 66 and Reference 67, and take notes related to intercultural communication.

Task 4 Analyzing the cultural phenomena in movies

➲ You will have 2 weeks to accomplish this task.

4.1 The class will be divided into 5 groups and each group will be assigned one cultural phenomenon in the table below.

Cultural phenomenon/ Group	Movies	Research questions
1 Racism	*Crash* *Men of Honor* *Green Book*	What is Emancipation Proclamation? What is Jim Crow Laws? How racism is explained in these movies? What are **the solutions to the (cross) cultural conflicts**?
2 Language varieties	*The Ladykillers* (2004)	What is African American English (AAE)? What are the other language varieties in this movie? Allusions form the Bible and literary work in this movie What are **the solutions to the (cross) cultural conflicts**?

(To be continued on next page)

(continued)

Cultural phenomenon/ Group	Movies	Research questions
3 **Melting Pot, Salad bowl or Mosaic**	*My Big Fat Greek Wedding* (Episode 1, 2002)	What does the word "big" mean? What does the word "fat" mean? What does the word "Greek" mean? How do you understand the speech Toula's father gives at her wedding reception? What are **the solutions to the (cross) cultural conflicts**?
4 **Cultural conflicts**	*Babel* *Joy Luck Club* *Lost in Translation* *Passage to India* *The Farewell* (2019)	What is the origin of the film name *Babel*? Why is the movie named *Babel*? What is *Joy Luck Club* meant by the author? Who is the author of this novel? How much do you know her? What is/are lost in translation? Who is the author of the novel *Passage to India*? How much do you know him? What is the background of this work? The movie *The Farewell* was released in China as *Don't Tell Her* (《别告诉她》). What triggers the conflict between the girl Billi and her family? What are **the solutions to the (cross) cultural conflicts**?

4.2 Watch the movie(s) in your topic area. The movies can be found on the video-share websites, e.g. Aiqiyi.

4.3 Do research on the cultural phenomenon and research questions for your group. Note down the information and make the matrix.

4.4 Do a panel discussion. (Please prepare this panel at least one week before the event.)

- **One culture phenomenon, one panel**

· **One moderator, one panel**

· **In presentation session, do PPT presentation; every presentation should include**

a. the information of the items or research questions in your topic area

b. your analysis of the relative cultural elements in the movie(s) with the evidence in the movie(s)

c. your ways to deal with the barriers in intercultural communication in the movies suppose you were in such situations. You may refer to the websites in Reference 68 and Reference 69.

☺ Repetition of information and points is not allowed in each panel, that is, every presentation in each panel should be given form different perspectives.

☺ Time limit: 5 – 8m's/person

· **In discussion session (15m's),**

a. you can ask and answer questions to have a further discussion about your topic area;

b. you can propose some solutions to the problems in intercultural communication;

c. you can mention the inspiration for your future international academic communication from this panel discussion.

Discussion session will start after all the presentations in each panel are finished.

Unit 2

Telling Chinese Stories Well

Learning objectives

- How to express our Chinese culture in cross-cultural communication
- How to make our voice heard in cross-cultural communication

Task 1 Learning to express Chinese culture

1.1 Work in a small group and have a discussion on the case.

Case

Suppose you are having your study abroad now. When dining with your professor and classmates, you want to share the Chinese food culture with them by talking about some classic traditional foods:

1. 包子
2. 饺子
3. 馄饨
4. 元宵
5. 汤圆
6. 粽子

Before your discussion, you have to know that it is widely accepted that the English term for "饺子" is "Chinese dumpling" and for "馄饨" is "Wanton".

- How can you express yourself clearly in this situation?
- What possible problems are there in your communication with your teacher and classmates?
- How can you solve the problems?
- According to this case, do you have any friendly reminders of expressing our Chinese culture in cross-cultural communication?

☺ Brainstorming or six-thinking hats may help your group discussion.

1.2 Present your discussion results before your classmates. While listening to other groups' presentation, note down the important information.

1.3 Work in your group and summarize the ideas in presentation related to the proper ways to express our Chinese culture in cross-cultural communication. Hand in your summary when your discussion is finished.

Task 2 Learning how to make our voice heard in cross-cultural communication

2.1 Watch the videos about traditional Chinese values in Reference 70.

While watching the videos, think about the questions:

1. What way (s) is (are) used to make the audience in other cultures understand the Chinese values?
2. Why is/are the way(s) are employed by the video-producer?
3. Do you think the way(s) are persuasive to the audience? Why?
4. Do you have different persuasive ways to express our Chinese traditional values? What are they?

☺ Note down your answers to the questions.

2.2 Do a Fish-tank discussion about the questions in 2.1.

· The whole class can be divided into 4 or 6 groups; 3 or 4 persons in each group.

· Two groups work together and play the roles of posters and lurkers in turns. That is, when one group is having their discussion, the other group is watching, listening and taking notes. When one group finish their discussion, change the roles.

· When the discussion is finished, two groups sit together and have the Q&A session.

· After the Q&A session, two groups will have the further discussion about another question: What should we do when we want to make the world hear our voice? While having your discussion, play the proper roles in the discussion.

☺ You may refer to the roles in 1.1, Task 1, Unit 2, Section 5 before this discussion.

Task 3 Wrapping it together

Work in a group of 3 or 4 members and prepare a pair presentation in a pop-sci forum:

- The presentation will be about one of the topics:

  羊肉泡馍

  糖蒜

  手擀面

  泡菜

- Your presentation should include at least two parts:

  a. Culture of the food

  b. Science in making or eating the food

- Length of the presentation: 15m's

References

1 https://graduate. universityofcalifornia. edu/admissions/applying/personal - statement/index. html

2 https://owl. purdue. edu/owl/job_search_writing/preparing_an_application/writing_the_personal_statement/index. html

3 https://www. petersons. com/blog/what - should - i - write - about - in - my - graduate - personal - statement/? _ga = 2. 186152493. 521874207. 1587570087 - 1982280915. 1547015944

4 https://admission. universityofcalifornia. edu/how - to - apply/applying - as - a - transfer/personal - insight - questions. html

5 https://owl. purdue. edu/owl/general_writing/graduate_school_applications/graduate_school_applications_statements_of_purpose/index. html

6 https://grad. berkeley. edu/admissions/apply/statement - purpose/

7 https://www. northeastern. edu/graduate/blog/how - to - write - a - statement - of - purpose/

8 http://www. essaylabb. com/sop - personal - statement - essay - writing - service - under - graduate. htm

9 https://www. kent. ac. uk/ces/protected/CV%20top%20tips. pdf

10 https://www. kent. ac. uk/ces/student/Example% 20Mar2013% 20skills - based%20CV. pdf

11 https://www. kent. ac. uk/ces/student/Example% 20skills - based% 20CV% 20NEW. pdf

12 https://career - advice. jobs. ac. uk/cv - and - cover - letter - advice/academic - cv - template/

13 https://open. lib. umn. edu/communication/chapter/11 - 3 - persuasive - reasoning - and - fallacies/

14 https://www. ets. org/gre/revised_general/prepare/analytical_writing/argument/scoring_guide

15 https://writingcenter. unc. edu/esl/resources/academic - integrity/

16 https://teaching. berkeley. edu/resources/design/academic - integrity

17 http://www. sussex. ac. uk/skillshub/? id = 287

18 https://style. mla. org/plagiarism - and - academic - dishonesty/

19 https://usingsources.fas.harvard.edu/writing-original-papers
20 https://usingsources.fas.harvard.edu/
21 https://www.sciencedirect.com/science/article/pii/S0020722517321535
22 https://www.researchgate.net/publication/49591443_A_Review_on_Rheological_Properties_and_Measurements_of_Dough_and_Gluten
23 https://d5qwk5ky11xwv.cloudfront.net/app/uploads/sites/3/2020/01/Gonzales_DisruptingWhiteNormativity.pdf
24 https://www.albany.edu/spatial/WebsiteFiles/ResearchAdvices/how-to-read-a-paper.pdf
25 http://www.health-policy-systems.com/content/10/1/36
26 onlinelibrary.wiley.com/doi/full/10.1111/j.1471-1842.2005.00549.x
27 https://ieeexplore.ieee.org/document/7911098
28 https://mcgraw.princeton.edu/sites/mcgraw/files/media/active-reading-strategies.pdf
29 https://www.scribbr.com/dissertation/literature-review/
30 https://owl.purdue.edu/owl/research_and_citation/conducting_research/evaluating_sources_of_information/evaluating_digital_sources.html
31 https://owl.purdue.edu/owl/research_and_citation/conducting_research/research_overview/synthesizing_sources.html
32 https://www.simplypsychology.org/synthesising.html
33 https://apastyle.apa.org/
34 https://apastyle.apa.org/style-grammar-guidelines/citations/quotations
35 https://www.yorksj.ac.uk/study/research/apply/examples-of-research-proposals/
36 https://advice.writing.utoronto.ca/types-of-writing/academic-proposal/
37 https://libguides.usc.edu/writingguide/researchproposal
38 https://www.scribbr.com/dissertation/research-proposal/
39 https://en.wikipedia.org/wiki/Research_question
40 https://writingcenter.gmu.edu/guides/how-to-write-a-research-question
41 https://libraries.indiana.edu/sites/default/files/Develop_a_Research_Question.pdf
42 https://www.monash.edu/rlo/research-writing-assignments/understanding-the-assignment/developing-research-questions

43 https://web. stanford. edu/class/polisci243c/ResearchProposalTemplate. html
44 https://www. scribbr. com/dissertation/research – proposal/
45 https://www. academia. edu/36292778/Title_of_research_Young_Childrens_Drawing_Behaviours_Supporting_Young_Children_Drawing
46 http://portal. unesco. org/en/files/1615/1019482428II11PromLitYouthAdEd. doc/II11PromLitYouthAdEd. doc
47 https://www. hamilton. edu/documents/Sample% 20Bio% 20Lab% 20Report. pdf
48 http://www. phrasebank. manchester. ac. uk/
49 https://library. leeds. ac. uk/info/14011/writing/106/academic_writing
50 https://www. sydney. edu. au/students/writing/planning – writing. html
51 https://academicguides. waldenu. edu/writingcenter/scholarlyvoice/first
52 https://writingcenter. unc. edu/tips – and – tools/should – i – use – i/
53 https://academic – englishuk. com/wp – content/uploads/2017/07/Signposting – 2017 – AEUK. pdf
54 https://academic – englishuk. com/wp – content/uploads/2017/04/Presentation – Phrases – AEUK. pdf
55 Presentation Skills – how to give an academic presentation at university (academic – englishuk. com)
56 https://www. wikihow. com/Make – a – Great – PowerPoint – Presentation
57 https://edubirdie. com/blog/presentation – topics
58 https://www. monash. edu/rlo/assignment – samples/science/poster – presentation
59 https://www. myamericannurse. com/how – to – create – an – effective – poster – presentation/
60 https://urc. ucdavis. edu/sites/g/files/dgvnsk3561/files/local _ resources/documents/pdf_documents/How_To_Make_an_Effective_Poster2. pdf
61 http://elss. elc. cityu. edu. hk/ELSS/Resource/Effective% 20Discussion% 20Skills/
62 https://student. unsw. edu. au/discussion – skills#: ~ : text = % 20These% 20skills% 20include% 3A% 20% 201% 20introducing% 20the, summarising% 20the% 20group% E2% 80% 99s% 20discussion% 20and% 20achievements. % 20More% 20
63 https://accendoreliability. com/discussion – skills/

64 http://cet. usc. edu/cet/wp - content/uploads/2016/12/Strategic - Questioning - Techniques - for - Discussion. docx
65 https://www. nature. com/scitable/topicpage/panel - discussions - 13909630
66 https://www. teachingenglish. org. uk/article/intercultural - communicative - competence
67 https://open. lib. umn. edu/communication/chapter/8 - 3 - intercultural - communication/
68 https://open. lib. umn. edu/communication/chapter/8 - 4 - intercultural - communication - competence/
69 https://warwick. ac. uk/fac/cross _ fac/globalpeople2/knowledgeexchange/frameworks/competency/
70 https://www. chinadaily. com. cn/culture/topics/searchingforkungfu

Video 1: https://writingcenter. unc. edu/tips - and - tools/why - we - cite/
Video 2: https://www. youtube. com/watch? v = z0VF3kh1ffk
Video 3: https://www. youtube. com/watch? v = 2jZDBz0qVtM
Video 4: https://www. cambridge. org/files/3115/2847/5405/Cambridge_Academic_English_Advanced_C1_Unit_A_Video. mp4
Video 5: https://www. youtube. com/watch? v = lpvgfmEU2Ck
Video 6: https://www. youtube. com/watch? v = I2xyxABvizM
Video 7: https://www. youtube. com/watch? v = kgkX2FHMbcE

Appendix Ⅰ

INTRODUCTION	· The reading and the lecture are both about __________, which is __________. (topic) · The author of the reading believes that __________. (mainidea) · The lecturer casts doubt on the claims made in the article. · He/She thinks that __________. (main idea)	*The reading and the lecture are both about ecocertification of wood products, which is a way to show that they are environmentally friendly. The author of thereading believes that American companies will not adopt this practice. The lecturercasts doubts on the claims made in the article. She thinks that American woodcompanies will eventually certify their products.*
1st BODY PARAGRAPH	· First of all, the author points out that ______. (point 1) · It is mentioned that __________. (supporting detail) · This point is challenged by the lecturer. · She says __________. (counter – point 1) · Furthermore, she argues ________. (supporting detail)	*First of all, the author points out that customers will likely ignore such a label. It ismentioned that many products are now given special labels, so shoppers no longertrust them. This point is challenged by the lecturer. She says customers actually dopay attention to claims when they are made by official agencies. Furthermore, sheargues that Americans will be enthusiastic about products that are endorsed by atrustworthy organization.*
2nd BODY PARAGRAPH	· Secondly, the author contends that __________. · The article notes __________. · The lecturer rebuts this argument. · She suggests __________. · She elaborates on this by mentioning that __________.	*Secondly, the author contends that it costs a lot to have wood inspected, so certifiedproducts will be more expensive. The article notes that American consumers arestrongly motivated by price, and will choose products that have not beencertified. The lecturer rebuts this argument. She suggests that customers do not caretoo much about small differences in price. She elaborates on this by mentioning thatcertified products will only be about five percent more expensive, which will not affectthe purchasing decisions of buyers.*

(To be continued on next page)

(continued)

3rd BODY PARAGRAPH	· Finally, the author states that ________. · The article establishes that ________. · The lecturer, on the other hand, posits that ________. · She puts forth the idea that ________.	*Finally, the author states that certification only makes sense for companies that sellproducts outside of the United States. The article establishes that American firmssell most of their products domestically, and their customers are happy with theirmerchandise as it is. The professor, on the other hand, posits that Americanbusinesses should be afraid of foreign competitors. She puts forth the idea that-foreign firms could flood into America and win customers by selling ecocertifiedwood to people who care about the environment.*
No *conclusion is needed.*		

Appendix Ⅱ

Memorandum *of Discussion*

To:

From (including names and roles in the discussion):

Date:

Subject:

Discussion strategies:
